REFORMING
THE MILITARY
RETIREMENT
SYSTEM

Beth J. Asch
Richard Johnson
John T. Warner

Prepared for the
Office of the Secretary of Defense

National Defense Research Institute

RAND

This report applies the models developed in earlier RAND reports (Asch and Warner [1994a, 1994b]) to analyze the effects of converting the military retirement system to a system very similar to the Federal Employees Retirement System (FERS), which is the system that covers federal civil service employees. The authors analyze several versions of this system, named the Military Federal Employees Retirement System, or MFERS. To compensate for the mandatory contributions, MFERS would be coupled with a pay raise. To the extent that the MFERS system failed to create the retention and separation patterns needed to achieve the services' desired seniority profiles, it would also include a system of retention bonuses and/or separation payments. The authors consider the implications of this proposal in terms of the effects on cost, force size and structure, productivity, and force management flexibility.

This research was conducted for the Director of Program, Analysis, and Evaluation of the Office of the Secretary of Defense within the Forces and Resources Policy Center in RAND's National Defense Research Institute, a federally funded research and development center sponsored by the Office of the Secretary of Defense, the unified commands, the Joint Staff, and the defense agencies.

The report should be of interest to policymakers concerned with the military retirement system and to others studying the career military or compensation issues in large organizations.

CONTENTS

FIGURES

TABLES

the-board pay increase to compensate members for mandatory contributions to the retirement plan, and the third part is a set of retention bonuses targeted to specific groups (such as occupations) to address any retention problems. MFERS would also consist of three parts: Social Security benefits, a defined benefit plan (called the basic plan) that vests employees at five years of service in an old-age annuity, and a defined contribution plan (called the thrift savings plan) that vests employees at three years of service and that matches employee contributions up to 5 percent of basic pay.

MFERS is a less-generous system than the current military retirement system for those who separate with 20 or more years of service. Consequently, we find that military retention falls significantly in the steady state under MFERS. Because the empirical version of our model is not occupation specific, it cannot easily accommodate the retention bonuses that initially were considered to address this problem. Therefore, to address this retention problem, we analyze in this report a system that couples MFERS with an active duty pay increase that is larger, on average, than the 7 percent increase included in the second part of the alternative plan. Although pay and bonuses are distinct policy options for addressing retention problems (for example, they have different cost implications), they are similar insofar as, together with MFERS, they both result in a compensation system that places a greater share of compensation in the form of active compensation and less in the form of retired pay, in contrast to the current system. Thus, for the purposes of much of our analysis, the plan we analyze (MFERS plus a pay raise) and the plan we initially set out to evaluate (MFERS plus retention bonuses) are similar. However, throughout this summary and the report, we note where our findings for the two plans might differ.

We first estimated the pay incentives necessary to maintain a force similar in structure to REDUX[1] and evaluated its productivity effects. We find that coupling MFERS with a 13 percent across-the-board pay raise is sufficient to produce the same force size and structure under MFERS as under REDUX. However, we also estimate that with an across-the-board pay raise, our measures of productivity would fall.

---

[1]REDUX is the current military system for those members who entered service after August 1, 1986.

Average effort is predicted to fall by about 2.5 percent and average E-9 ability—our measure of ability sorting—is predicted to decline by 21 percent. The productivity measures are predicted to decline because MFERS reduces the main source of deferred compensation under the current compensation system—retired pay—and thus the main source of effort and ability sorting incentives. An across-the-board pay raise that gives the same percentage raise to all individuals does not offset this reduction in deferred compensation. In other words, MFERS with an across-the-board pay raise undoes much of the skewness of the current compensation system. Therefore, although MFERS with an across-the-board pay raise can maintain the force size, we predict that it is not an improvement over the current system (given that force size and structure are being held constant) because our measures of productivity are predicted to decline.

We find that for MFERS to be an improvement over the current retirement system, it must be coupled with a skewed pay raise—higher raises in higher grades (or, alternatively, a skewed set of retention bonuses). Coupling MFERS with a skewed pay raise can increase productivity (and reduce costs) relative to REDUX while producing the same general force size and structure. Specifically, for the Army enlisted force, we estimate that average effort rises by 17 percent and average E-9 ability rises by 87 percent.

For MFERS to represent an unambiguous improvement over REDUX, it must reduce costs at the same time it maintains the force structure and raises productivity. Costs are composed of active duty pay plus an accrual charge to fund future retirement liabilities of the current force. A critical element in costing is the real discount rate used to determine the military retirement accrual charge. Until very recently, the Department of Defense (DoD) Actuary used a 2 percent real rate in estimating the accrual charge. Beginning in FY 1995, the Actuary raised its real discount rate assumption to 2.75 percent. The real interest rate is an important determinant of the cost of the military retirement system, or the savings from changing it. An increase in the real discount rate reduces the accrual charge for the current force and tends to reduce the savings to be had from implementing policy changes that reduce future retirement outlays.

Just what the real discount rate for public decisions should be is an open question. Although the DoD Actuary raised its discount rate assumption in FY 1995, 2.75 percent is still a relatively low real discount rate. In fact, there is a substantial body of economic literature arguing that the real discount rate that should be applied to public decisions is even higher than 2.75 percent. We account for the uncertainty in real discount rates by evaluating the cost of REDUX and MFERS under several alternative assumptions about the real government discount rate. We find that when 2 percent is used to calculate the accrual costs of either retirement system, MFERS with a skewed pay raise would reduce total manpower costs by about 6 percent and result in annual savings to DoD of about $2.4 billion based on FY 1997 force levels. When the government discount rate is very low, MFERS appears to be a clear improvement over REDUX on grounds of both higher productivity and reduced cost.

The case for MFERS is less compelling the higher the real discount rate. When the real discount rate is increased to 2.75 percent, the savings in total manpower costs decline to 2.2 percent (about $1 billion for the 1997 force). At this discount rate, MFERS may still be an improvement in that it raises productivity while costing slightly less. But when the discount rate is increased to 5 percent, MFERS is estimated to cost about 6 percent more than REDUX. In this case, the case for MFERS would depend on whether the estimated improvements to productivity are worth the extra costs, something that policymakers would have to decide.

We also used our empirical model to predict the pattern of retention and costs in transition to the steady state under two cases (assuming a 2 percent real discount rate). In the first case, current members would be grandfathered under REDUX and new entrants would be automatically enrolled into MFERS with a skewed pay raise. In the second case, current members would be allowed to convert to the new system. We find that although there is some variation in retention around the steady-state level, the variation is not large. We also find that most of the cost savings associated with moving to the new system would occur in the first three years in both cases. Beyond the third year, there are small variations in cost around the steady state. Thus, our model predicts that there would be no large spikes in retention or cost, and most of the cost savings (under a 2 percent real discount rate) would occur fairly soon under both cases.

The empirical model evaluates the productivity and costs of a force under MFERS that is comparable in rank and experience structure to the one produced by REDUX. MFERS has several implications for force management that are not reflected in the productivity and cost estimates discussed above. The 20-year retirement system creates a similar experience profile across the broad spectrum of military skills (Asch and Warner [1994b], Table 2). It has been argued that because of the "lock-in" effect created by the 20-year system, the services have little flexibility to alter the experience distribution of their forces even when productivity considerations might merit such changes. By diminishing the influence of retired pay in retention decisions of junior and mid-career personnel and by permitting more flexibility in the use of active duty pay (e.g., through bonuses), MFERS offers the potential for more flexibility in force management. At the individual level, MFERS could provide the services with a means of separating marginal performers in mid-career whom they are now reluctant to separate prior to the 20-year mark.

Although a more up-front compensation system—which MFERS coupled with a skewed active pay increase or greater use of bonuses would produce—would permit greater flexibility of force management at the junior or mid-career levels, MFERS might create inflexibilities of its own in the management of senior personnel. The current 20-year system offers strong incentives for senior personnel to leave upon qualifying for retirement benefits. But under MFERS, coupling skewed pay increases with a reduction in immediate separation benefits creates stronger retention incentives for senior, high-ranking personnel. The result would be higher retention among the personnel who remain to the 20-year mark and beyond. Although the increased numbers of senior personnel might be beneficial in many skill areas (e.g., the medical corps), superannuation might be a problem in skills demanding youth and vigor. If senior personnel are unwilling to leave voluntarily, the services would have to rely more heavily on involuntary separation to maintain the youth and vigor of the force than they do under the 20-year system with its inducement to senior personnel to separate voluntarily.

The prospect of a heavier reliance on involuntary separation of senior personnel to maintain the experience distribution of the force might impose what we have termed "organizational influence costs" (Asch and Warner [1994a]). These costs include the potential for lower morale among senior personnel who are faced with the prospect of involuntary separation or the lower productivity that might be associated with modification of personnel policies to permit senior personnel to stay for longer careers. These organizational influence costs are difficult to measure but could swamp the savings associated with either version of MFERS.

While it is ambiguous whether MFERS offers improvements in force management over REDUX, MFERS does offer personnel an important advantage over REDUX—portability. MFERS would allow military members to transfer their accumulations and vested benefits to the civil service system, FERS. Although the evidence is limited on the number of military members who enter the civil service after their military service, anecdotal evidence suggests that the number may be high.

Finally, because the current back-loaded system creates strong incentives for career personnel to stay, it creates force stability and prevents premature losses in the face of uncertainty such as unforeseeable fluctuations in the labor market. How would MFERS compare on this dimension? Since MFERS more resembles a front-loaded compensation system, it would seem to compare unfavorably. However, if MFERS is coupled with a skewed pay raise, much of the member's compensation would still be deferred, and this deferred compensation would help create stability among the career force. Even if MFERS is coupled with an across-the-board pay raise, it might give added protection against retention fluctuations among more junior personnel. Therefore, whether MFERS creates more force stability relative to the current system is unclear.

Given these considerations together with the results of our empirical model, the question is, should the military adopt either plan? We cannot say, because the answer depends on how policymakers weigh the advantages and disadvantages of the systems. However, there is an alternative proposal that would maintain the advantages of the two systems while addressing their disadvantages. This alternative would include three components: MFERS, a skewed pay raise, and a system of cash separation pays.

Exercising our empirical model (and assuming a 2 percent real discount rate), we find that the alternative proposal would reduce costs and increase our measures of productivity relative to REDUX while producing the same general force size and structure. It would also be portable. Most important, because this system includes separation pay, it would most likely reduce the organizational influence costs associated with involuntary separation. In essence, like the Voluntary Separation Incentive/Special Separation Benefit (VSI/SSB) program used to facilitate the drawdown, the separation pay under this system would ease the transition of individuals who must leave the service. Since the transition of these individuals would be smoother, the services would likely be more willing to separate these individuals, and the system would thus enhance force management flexibility. A disadvantage of this proposal is the risk that the separation pay would be operated like a bonus program that is subject to frequent changes. Because such frequent changes would create uncertainty about benefits and have adverse effects on behavior, once the separation pay scheme was in place, the formula and target populations should change rarely. If this disadvantage could be overcome, then this system would likely be an improvement over the current system as well as the two systems analyzed in this report.

ACKNOWLEDGMENTS

There are several individuals to whom we owe a debt of gratitude. We would like to thank our project officer, Russ Beland, for his comments on a earlier draft, as well as Brian Jack and Craig College within the Office of the Director of Program, Analysis, and Evaluation for their input. We are indebted to Saul Pleeter for providing several necessary data inputs as well as for his long-term support for our work on military compensation issues. Within RAND, we benefited from numerous conversations with and formal reviews from Glenn Gotz and James Hosek. These conversations and reviews helped clarify our thinking about the MFERS alternative. We also benefited from the comments of Susan Hosek on an earlier draft. All errors, of course, remain our own.

Chapter One

INTRODUCTION

The current military retirement system dates back to 1947, when Congress implemented a common system for the military services and for officers and enlisted personnel alike. As a result of modifications in 1981 and 1986, there are actually three systems now in effect,[1] but the basic structure has not changed: the system provides an immediate lifetime annuity to those who separate with 20 or more years of active duty, but no benefits to those who separate with less than 20 years (unless those separatees subsequently participate in the reserves long enough to qualify for a reserve pension beginning at age 60).

---

[1]The three systems are structured as follows. Pre-FY 1981 entrants receive retired pay according to the formula .025*YOS*final basic pay (where YOS denotes years of service), such that 20-year retirees receive 50 percent of final basic pay and 30-year retirees receive 75 percent. Retired pay for this group is fully protected from inflation. Retired pay for those who entered between FY 1981 and FY 1986 is calculated similarly except that pay is based on the individual's high three years' average basic pay (high-3) rather than final basic pay. It is also fully indexed for inflation.

The Military Retirement Reform Act of 1986, also known as REDUX, implemented several important changes. First, the annuity formula was changed to [.40 + .035*(YOS − 20)]*high-3 average basic pay for the years between separation and age 62, at which time pay reverts to .025*YOS*high-3 average basic pay. Consequently, retired pay during the transition between military service and full retirement ranges between 40 percent of high three years' average basic pay at YOS 20 and 75 percent of high three years' basic pay at YOS 30. Second, rather than indexing retired pay for inflation, the annual cost-of-living adjustment (COLA) between separation and age 62 is 1 percent less than the percentage growth in the Consumer Price Index (CPI). At age 62, retired pay is then fully adjusted for the CPI growth since separation. Thereafter, it again increases according to the CPI-minus-1-percent rule. The 1986 reforms thus changed the system by (1) reducing the amount received at YOS 20, (2) raising the growth in retired pay for each year served after YOS 20, and (3) reducing the real value of the stream of retired pay in an inflationary environment.

From the start, the military retirement system has been the target of critical analyses by Department of Defense (DoD) study groups, presidential and congressional commissions, and independent analysts.[2] Various critics have charged that the system is: (1) excessively costly and unfair to taxpayers, (2) unfair to the vast majority of military entrants who do not serve long enough to receive retirement benefits, (3) inefficient, and (4) inflexible.

To the general public, the two most visible aspects of the system are its cost and the relatively young ages of military retirees. The fiscal year (FY) 1994 retired pay accrual for active duty personnel was $11.7 billion. A noted defense analyst, Jacques Gansler, has noted that "The military retirement program, though politically loaded, is likely to be forced to change because of cost considerations." Further, "more and more people have been retiring at about 40 years of age, depriving the services of their expertise and collecting retired pay for the rest of their lives."[3] The implication here is that retirees are departing before the services would like for them to and are receiving "excessive" benefits at the expense of taxpayers.

Other critics charge that it is unfair for 20-year separatees to receive a lifetime retirement annuity while others who serve for shorter periods receive nothing. The fact that only some 30 to 40 percent of officer entrants and 10 to 15 percent of enlisted entrants will stay for a full 20-year career and receive benefits is seen to be unfair to those who receive no benefits for time served. The military, in fact, is one of the few organizations exempted from the Employee Retirement Income Security Act (ERISA), the federal law that requires private sector employers to vest employees in their retirement systems after (usually) five years of service. Some have argued that the military should be brought under ERISA's early vesting requirements.

---

[2] See, for example, the reports of First and Fifth Quadrennial Reviews of Military Compensation conducted by DoD, the report of the Defense Manpower Commission (1976), the *Report of the President's Commission on Military Compensation* (1978), and the *Final Report of the President's Private-Sector Commission on Government Management* (1985).

[3] See Gansler (1989, pp. 297–298). Although Gansler wrote in 1989, the REDUX system implemented in August 1986 in fact substantially reduced benefits for those who separate with 20 years of service and has added incentives to serve beyond the 20-year mark.

Implicit in the charge of "excessive cost" is the belief that military forces of the same quality could be obtained more cheaply. In other words, the system is inefficient. Evidence suggests that personnel discount future dollars at a much higher rate than the government's borrowing rate. It therefore follows that the same force could be obtained at lower overall cost with more reliance on "up-front" (active duty) pay and less reliance on retired pay. Taken to its extreme, the "up-front" view says that there need be no retirement system at all: the most efficient compensation system is an active-pay-only system. Of course, some advocates of this line of reasoning recognize that it would be politically infeasible to eliminate the system altogether and therefore recommend a less-generous system that conforms to ERISA guidelines for private sector pension plans.

Critics also charge that the military retirement system inhibits force management flexibility. The services are well aware of the financial costs imposed on mid-careerists who are involuntarily separated prior to the 20-year vesting point. As a result, beyond a certain grade or YOS, personnel are treated as if they have an implicit contract. The services are reluctant to separate all but the poorest performers for fear of the effect of involuntary separations on morale. That is, the services' "desired" force structures reflect the actual retention patterns that emerge as a result of the current compensation system. Without the constraint of the current retirement system, the "desired" force may differ significantly.

In response to the various criticisms levied against the retirement system, numerous changes to the retirement system have been proposed. In this report, we evaluate one alternative that would convert the current military retirement system to a system similar to that covering federal civil service employees—the Federal Employees Retirement System, or FERS. We call this retirement plan MFERS (Military Federal Employees Retirement System).[4] Since MFERS would require members to contribute to their retirement plan, it would be coupled with a 7 percent across-the-board pay raise to compensate members for these mandatory contributions and their tax implications. A system of retention bonuses would also be im-

---

[4]As discussed in Chapter Two, MFERS differs from FERS in several minor ways. Asch and Warner (1994b) analyzes other options for changing the military retirement system.

plemented to offset any retention problems the services might experience under this plan.

Since MFERS would vest workers significantly earlier than the current 20-year system (as early as after three years of service), it would offer benefits to military personnel who are ineligible for benefits under the current system. Thus, on face, MFERS would appear to address the criticism that the current system is unfair to those who separate before year 20. MFERS would also decrease the expected value of retirement benefits to military members since it is a less valuable plan. To keep military personnel equally well off under MFERS relative to the current system given that members must contribute to the retirement plan, MFERS would be coupled with a 7 percent across-the-board pay raise and a system of retention bonuses. By converting the military system over to MFERS, the military compensation would be more "front-loaded"—a greater proportion would be up-front in the form of active duty pay and less would be deferred in the form of retired pay. This front-loaded system would reduce costs, and thus would seem to address the criticisms that the current system is excessively costly and inefficient.

To address criticisms about the appropriateness and efficiency of the military retirement system and to evaluate alternative systems such as MFERS, a theory or model is needed that recognizes the military's manpower goals, incorporates the essential features of the military organization, and predicts the behavioral responses of personnel to alternative compensation and personnel policy. Until recently, such models have narrowly focused on the relationship between compensation and retention behavior and the resulting years of service structure of the force and have ignored the other consequences of the military's personnel and compensation system. In particular, less attention has been paid to questions of productivity: (1) whether the system induces the most able personnel to stay and seek advancement to the highest ranks, and (2) whether the system encourages personnel to work hard and effectively.

In two earlier reports, we developed a theoretical model that allows an analysis of these issues (Asch and Warner [1994a]), and an empirical version of that model (Asch and Warner [1994b]) that we can then use to analyze various military retirement system reform proposals. In this report, we use the empirical model and the theoretical insights derived from the theoretical model to analyze the implications of converting the military retirement system to MFERS. Since our empirical model can not easily accommodate retention bonuses, we analyze the case in which MFERS is coupled with, on average, a pay raise higher than 7 percent and no bonuses.[5] Our focus is on the implications for active duty personnel. We derive results not only in the steady state—when all members would be under the new system—but in transition to the steady state. We consider two alternative strategies for transitioning to the steady state. In particular, we consider the case where all current members are "grandfathered" (i.e., would remain) under the current system(s) and only new members would participate in MFERS, and the case where current members would be allowed to voluntarily convert to MFERS.

Chapter Two discusses the MFERS alternative in detail. Chapter Three presents an overview of the theoretical and empirical models developed in our earlier papers. Chapter Four presents our steady-state results, and Chapter Five presents our results relating to the transition to the steady state. In Chapter Six, we discuss other factors that may be important in considering whether to convert the military system over to MFERS. Policy implications and conclusions are discussed in Chapter Seven.

---

[5]Essentially, we are addressing any retention problems created by MFERS with a pay raise rather than with bonuses. As discussed in Chapter Four, pay and bonuses differ in several respects, but are similar to the extent that using them results in a more front-loaded compensation system.

Chapter Two

OVERVIEW OF MFERS

In this chapter, we describe the compensation scheme we evaluated. This scheme consists of a retirement system which we call MFERS, a pay raise, and a system of retention bonuses. MFERS consists of three main parts that are identical to the three main parts of FERS: (1) a defined benefit plan called the "basic benefit plan" where the member's benefit is predetermined by a formula, (2) a defined contribution plan called the "thrift savings plan" where the member's benefit is determined by market forces and where workers have several withdrawal options should they separate from federal service, and (3) Social Security benefits.

We describe below these three parts of MFERS in greater detail, highlighting the vesting provisions, benefit determination methods, and other key provisions, as well as the ways in which MFERS would differ from FERS. We also describe two methods that could be adopted to transition the current system to MFERS. One is simply to grandfather current members under the new system, as was done for prior retirement system changes. The second would allow current members to convert to MFERS.

BASIC BENEFIT PLAN

Just as with FERS, the basic benefit plan under MFERS would vest members at five years of service. That is, a member must have at least five years of service to be eligible to receive benefits under the plan. Under the basic benefit plan, members would be required to contribute a small percentage of their basic pay—7 percent minus the tax rate under Social Security's OASDI (Old-Age, Survivors, and

Disability Insurance) program. In 1994, this percentage was .8 percent. However, the member would have the option of getting a refund of these contributions with interest instead of the retirement annuity when he or she left federal service.

The benefit formula under the basic plan equals 1 percent of an individual's highest three-year average pay times the years of service (YOS). If the member retires at or beyond age 62 with 20 or more years of service, the formula is 1.1 percent of highest three-year average pay times YOS. The normal age of retirement (i.e., the age when an individual can leave the service and begin collecting this benefit) depends on the member's years of service. The schedule is shown in Table 1. For those with five years of service, the normal retirement age is 62. For those with 20 YOS, the normal age is 60. For those with 30 YOS, the normal age is between 55 and 57, depending on one's date of birth.

The basic benefit plan also allows for early retirement. Those who have 10 years of service could retire as early as age 55 or age 57, depending on their date of birth. The basic benefit plan also would give a cost-of-living-adjustment (COLA) to those age 62 and older. This adjustment would equal the change in the Consumer Price Index (CPI) if the change in the CPI is less than or equal to 2 percent. It would equal 2 percent if the change in the CPI is between 2 and 3 percent and would equal the percentage change in the CPI minus 1 percent if the change in the CPI exceeds 3 percent.

Table 1

**Normal Age of Retirement
Under Basic Benefit Plan**

Age	Years of Service
62	5
60	20
55-57[a]	30

[a]The allowed normal retirement age depends on date of birth.

THRIFT SAVINGS PLAN

The thrift savings plan is a defined contribution plan that shares many features with the 401(k) pension plans found in the private sector. Under the plan, the government makes automatic and matching contributions to a fund and the employee has several options for investing the fund, including investing it in a government securities fund, a common stock fund, or a fixed-income index fund (or some combination of the three). The government's automatic contribution for each member is 1 percent of the member's basic pay, in which the employee is vested after three years. The government will also match contributions made by the employee up to 5 percent of basic pay.[1] Individuals are immediately vested in their own contributions, and their contributions (and earnings from the contributions) are tax deferred.

The thrift savings plan has several provisions for withdrawing funds. Members can withdraw the balance of their account only if they leave federal service. The withdrawal options depend on whether the member is eligible for retirement benefits under the basic benefits plan, as determined by his or her age and completed years of service. If the member is ineligible and separates from federal service, he or she must transfer the vested account balance of the thrift savings plan to an Individual Retirement Account (IRA) or other eligible retirement plan.[2]

If the member is eligible for retirement benefits, he or she has three withdrawal options. First, the member can transfer the account balance to an IRA. Second, he or she can receive a cash lump sum or a series of equal payments. Finally, he or she can purchase a life annuity that can begin at the date of separation or later. If the member chooses the first option, he or she faces a 10 percent penalty for withdrawing from the IRA before age 59.5. Under option 2, if the

---

[1]Specifically, the government matches 100 percent of the employee's contribution for the first 3 percent; 50 percent of the employee's contributions for the next 2 percent; 0 percent of the employee's contributions above 5 percent. An individual can contribute a maximum of 10 percent of basic pay each period subject to Internal Revenue Service restrictions.

[2]This is true if the account balance exceeds $3500. If the balance is $3500 or less, the member receives an immediate lump-sum cash payment.

member receives any proceeds before age 55, there is a penalty equal to 10 percent of the amount received before age 59.5.

SOCIAL SECURITY BENEFITS

Both the current military compensation system and MFERS include Social Security benefits, and members' earnings under both plans are subject to Social Security taxes. Thus, with regard to Social Security benefits, there is no difference between MFERS and the current military system. The only difference that can arise is if MFERS is coupled with an active duty pay increase. In this case, the members' Social Security tax liability is greater (since the tax is figured as a percent of earnings) and the members' future Social Security benefit is greater under MFERS since active duty pay would be greater. The net effect would depend on the rate of return on Social Security taxes. In the analysis in Chapters Four and Five, we ignore the differences in Social Security benefits and liabilities between MFERS and the current military retirement system.

COUPLING MFERS WITH A PAY RAISE

Under MFERS, military members would be required to contribute to their basic benefit plan and could voluntarily contribute to the thrift saving plan, as discussed earlier. Under REDUX, the current system, members make no contributions to retirement. To reimburse members for their mandatory contributions under MFERS's basic plan and the tax consequences of those contributions, MFERS would be coupled with a 7 percent pay increase. Any retention problems that occurred under this system would be addressed through retention bonuses that could be targeted to distinct populations.

DIFFERENCES BETWEEN MFERS AND FERS

MFERS differs from FERS in several respects. First, MFERS would allow medical, Commissary, and Exchange benefits for those military personnel who reach 20 years of service, just as under the current military retirement system. Second, the thrift savings plan under FERS features a loan program. Individuals can borrow from their contributions to the plan for the purchase of a home, educational expenses, medical expenses, or because of financial hardship. The ability to cash out contributions based on financial hardship when leaving military service would be made explicit under MFERS. We ignore these features of MFERS in our analysis in this report and focus solely on the basic plan and the thrift savings plan features.

TRANSITIONING TO MFERS

One transition option is to grandfather current members under REDUX; another is to allow current members to voluntarily convert to MFERS or remain under REDUX. Obviously, those who convert would forgo the expected benefits they would have earned from the current military retirement system. Furthermore, their previous years of service under the current system would not be included in the calculation of benefits under MFERS. Thus, for example, members who convert to MFERS at 10 years of service would have only one year of service under MFERS when their YOS was 11.

To compensate members who convert to MFERS for the loss in expected retirement benefits that would occur by moving away from the current system, those who convert would earn double the government's thrift savings plan contributions for a period of time equal to years of prior service. For example, a member with six years of service who converts to MFERS and contributes 5 percent of basic pay into the thrift savings plan would earn DoD contributions equal to 10 percent of basic pay for the next six years. This would allow converting members to gain some credit for their prior years of service. Nonconverting members would also have an opportunity to contribute to the thrift savings plan, but none of their contributions would be matched by DoD.

Chapter Three

OVERVIEW OF THE THEORETICAL AND EMPIRICAL MODELS

This chapter presents an overview of the theoretical and empirical models that underlie the results presented in Chapters Four and Five. In the discussion of the theoretical model, we summarize the military's objectives, the factors influencing individual decisionmaking, and the organizational policies that are most relevant for our analysis of the military retirement system. In the discussion of the empirical model, we describe the assumptions and general methods we used to generate the results presented in the next two chapters. A more formal presentation is given in Asch and Warner (1994a).

THEORETICAL MODEL

Organizational Objectives

The military's stated manpower goal is to attract and retain personnel in sufficient numbers to meet its grade and experience requirements. We call this the "macro" goal. Not so well recognized are several "micro" goals. First, personnel must be motivated to work hard and effectively. Since individual effort cannot be directly observed, compensation and personnel policies must be designed to provide individuals with the proper incentives to work hard and seek advancement. Second, the system must sort personnel effectively. That is, it must induce the proper person/rank/job matches, which requires retaining and promoting the more able to the higher ranks. Two implications follow. One is that low ability/effort individuals should be induced to leave. Another is that "climbing" (seeking ranks for which one is unqualified) and "slumming" (the converse of climbing) should also be discouraged.

Furthermore, given their hierarchical rank structures, the services want personnel to stay long enough to get a return on their training and experience, but not to stay too long. There must be enough turnover in the upper ranks to provide promotion opportunities for those in the lower ranks. Retention can be excessive, even among very able personnel. Consequently, the compensation system must be structured not only to provide the proper retention and effort incentives, but also to provide the incentive for personnel to separate when it is in the services' best interest for them to do so.

Military personnel managers have a variety of policy tools at their discretion. Compensation policy instruments that we consider include: (1) the level of entry pay, (2) the sequencing of promotion and longevity increases thereafter (i.e., intergrade and intragrade pay spreads), (3) bonuses and other skill-specific pay, and (4) the retired pay system. Personnel policy levers include minimum standards for retention and promotion and use of up-or-out rules. How do individuals respond to these tools? We address this issue below, but first we discuss some of our assumptions about individual productivity.

Individual Productivity

Past research shows that military recruits vary with respect to both their ability to perform tasks within the military organization and their "tastes" for military life. Despite the substantial sums spent screening new recruits, the military cannot perfectly measure entrants' true abilities. Rather, ability is revealed slowly over time. Nor can individuals' tastes be observed. We can only discern from unfolding retention decisions that stayers have stronger tastes for service than do nonstayers.

We can also assume that the military organization has difficulty monitoring individuals' work efforts. Although the military monitors work effort, it cannot do so directly or costlessly. While effort improves individual productivity, it also involves a cost—hard work. We assume that individuals do not like to exert work effort and would prefer to shirk if they could get away with it. In economists' terms, the marginal disutility of effort is positive.

We also assume that ability has a bigger impact on individual productivity in the upper ranks than in the lower ranks. That is, an individual with a low mental aptitude and one with a high mental aptitude may perform low-level tasks equally well, but the high-aptitude person is likely to make a better colonel or master sergeant than the person with low aptitude. Since higher-ranking personnel control more of the organization's resources and make decisions that have greater overall impact, it is important to have the most able personnel fill the upper slots. Individual work effort may also be more important at higher levels.[1]

Individual Decisionmaking

Of the variety of policy tools military personnel managers have at their disposal, the optimal policies will depend on individual decisionmaking. Once we understand how people behave and what factors influence them, policymakers can design policies to influence behavior according to the organization's goals.

Why do individuals join (or stay in) the military? We hypothesize that individuals join if they are better off doing so (in economic terms, if the expected utility from joining exceeds the expected utility from remaining in the civilian sector). The net payoff to joining depends partly on how long the individual remains in the military. Some join for only one tour, others for a 20-year career. We thus hypothesize that when deciding whether to join, individuals evaluate the payoffs to all the possible career paths that they might follow and weight each path by the probability that they will follow it. Career paths have dimensions of rank and years of service. Individuals with a lower taste for the military anticipate that they will not likely reenlist after an initial term. In contrast, individuals with stronger tastes for military life expect to serve longer, so they will place more weight on the payoffs associated with longer careers (e.g., retirement

---

[1]As Willis and Rosen (1979) discuss, a complicating factor is that ability is not unidimensional. Ability traits that are important for success in the lower ranks (e.g., physical strength or the capacity to follow orders) may not be the same as those required at the upper ranks (e.g., analytical reasoning or leadership skills). Skills that make one a good captain may not make one a good colonel. If this is the case, performance in the lower ranks may not be a good forecast of one's probable performance in the upper ranks, making selection for promotion more difficult. The problem is likely to be more severe in the officer ranks and it leads the services to stretch out the selection of officers for the senior ranks over time.

benefits). The benefits provided during the initial enlistment will dominate to a greater extent the enlistment decisions of low-taste personnel.

Aside from tastes, the decision to join depends in large part on the level of entry pay and its subsequent growth, with respect to both rank and longevity. Other important factors in the initial enlistment decision include the value of training received (especially its transferability to the civilian market) and educational benefits. An implication of our model is that ability has an ambiguous influence on the decision to join. If the more able have a higher expected payoff to joining (through, say, more rapid or more certain promotion or qualification for better educational benefits), they will be more likely to enlist. But the more able also have better civilian sector opportunities, which makes them less likely to join.

The decision to remain at each retention decision point thereafter is conceptually similar to the initial enlistment decision. Individuals are assumed to calculate the expected utility from remaining in the service by evaluating the payoffs to all possible future career paths and weighting the various paths by their probabilities. They will compare this utility with the utility from leaving immediately and stay if they expect to be better off. Again, we predict that high-taste individuals are more likely to stay, although more-able people may be more or less likely to stay than less-able people, depending on how ability is rewarded in the external market relative to the "internal" (service) market. The internal reward to ability depends in part on the extent to which the promotion system identifies and promotes the more able more rapidly and with higher probability. Even prior to the actual separation point, up-or-out rules induce separations of some personnel who know that they are likely to be affected by such rules.

Another factor that plays a role in retention decisions is the rate at which personnel discount future income. Research indicates that personnel have real discount rates in excess of 10 percent, as evidence from the drawdown separation program seems to confirm (see Warner and Pleeter [1995]). In our model, high discount rates serve to reduce the value of future pay relative to current pay and therefore cause individuals to place more weight on near-term pay in both their effort and retention decisions.

In addition to making retention decisions, personnel make choices about how hard to work. Individuals in our model supply effort in each grade and year of service up to the point where the extra (marginal) benefit of doing so equals the extra (marginal) cost. What factors affect effort? The answer is any factor that affects the marginal return or cost of effort. First and foremost in the military system is the return to promotion. Promotion to a higher rank provides a monetary reward and it may also yield psychological benefits. If future promotions depend on current performance, we predict that a higher monetary reward to future promotions should induce individuals to work harder in their current rank. The model also predicts that individuals will work harder in their current rank the more they value the status associated with higher rank. Similarly, individuals will work harder if doing so results in better future assignments, another nonpecuniary reward. Significantly, monetary rewards can come either through the active duty pay associated with higher rank or in the form of retirement benefits. Finally, individuals may also work harder in their current rank if there is an intragrade payoff that is contingent on effort. Performance bonuses or other nonmonetary rewards to top performers are hypothesized to spur effort.[2]

The military's hierarchical rank structure and the structure of its promotion contests affect effort expended. Subject to individual qualifications, personnel are promoted through the lower ranks with virtual certainty based on time-in-grade or time-in-service requirements. But beyond the junior ranks promotions are determined in competitive "contests" or "tournaments" in which only a fraction of those seeking advancement are promoted.[3] Competition at the up-

---

[2]Two caveats regarding our modeling of effort are in order. First, we do not explicitly model the effect of a member's effort on unit performance other than to assume that it has a positive effect. Thus, our model cannot tell us how much military output would rise when members' effort supply rises. Second, we assume for simplicity that a member receives no satisfaction from his or her unit's performance. Thus, our model does not account for the possibility that the member's satisfaction with service life may increase (which would cause his or her retention probability to rise) when unit performance improves as a result of an increase in other members' effort levels.

[3]Officer and enlisted promotion processes do differ somewhat. Officers are selected for promotion by selection boards and are promoted by entry year group. Failure to be selected within a specified YOS zone usually means the officer will never be promoted. Prior to the two highest grades, enlisted personnel are promoted on the basis

per ranks gets keener as a result of the declining fraction to be promoted and the increasing homogeneity of the pool of contestants.[4]

Some theoretical propositions follow. If the interrank pay spread is held constant, a declining probability of promotion tends to diminish work effort because personnel discount the reward to promotion by the probability that the reward will be received.[5] If the probability of promotion is low, individuals will not expend much effort to be promoted without a sufficient reward for promotion. Therefore, to maintain effort incentives with declining promotion rates, increasing interrank pay differentials are required.

The rate at which promotion chances improve with effort is also predicted to affect effort expended. Individuals are likely to work harder when extra effort improves their promotion chances a lot than when it improves them only a little. When the probability of promotion is very high, as in the junior ranks, individuals need not exert a lot of effort to ensure that they surpass the promotion threshold, so that the effect of effort on the probability of promotion is small. Likewise, when the probability of promotion is low, a change in effort may not improve one's promotion chance much. Therefore, marginal effort has little impact on the chance for promotion when the probability of promotion is either very high or very low. Extra effort has the most effect when the probability of promotion is around 50 percent.

In addition to the base promotion chance, the rate at which the likelihood of promotion improves with effort depends on the relative importance of random factors ("noise" or "luck") in the promotion contest. Because promotion in the lower ranks is based on explicit criteria or standards, luck has only a small influence on promotion outcomes. Luck assumes a larger role as individuals progress through the upper ranks. Having the "right" assignment, working for

---

of point systems and may accumulate the points required for promotion over a wide YOS range.

[4]Historically, the promotion rates to grades O-4, O-5, and O-6 have been around 80 percent, 70 percent, and 50 percent, respectively. Promotion rates to these ranks have declined considerably during the drawdown.

[5]This statement is technically true only if the probability of promotion is less than .5. More generally, as discussed later in the text, the effect of the probability of promotion on effort depends on the level of the promotion probability.

the "right" mentor, etc., loom larger as one progresses to higher levels. The increasingly more important role of luck serves to blunt the relationship between effort and the likelihood of promotion and thereby discourages effort as individuals progress through the ranks, all else equal.

The relationship between effort and the likelihood of promotion is also related to the composition of the promotion pool. In the lower ranks there is likely to be a lot of variation, or heterogeneity, in the skills and qualifications of those available for promotion. When the promotion pool is heterogeneous, it is easy for an individual to bypass some of the others by working harder. As individuals progress through the ranks, the pool available for promotion to the next rank becomes more homogeneous. Bypassing one's competitors by working harder becomes increasingly difficult the more alike are the individuals in the promotion pool. The increasing homogeneity of the individuals in the promotion pool is predicted to further blunt the relationship between effort and the likelihood of promotion.

Tastes and personal discount rates are also predicted to influence effort in the model. Individuals with a high taste for the military are more likely to stay for future periods and are thus more likely to reap the benefits of harder work today. Therefore, individuals with a high taste will work harder.[6] An important policy implication follows. Since first-termers have lower tastes than careerists on average, a pay raise targeted at the first-term force will not produce as much extra effort as a raise targeted at the career force. This result provides a rationale for skewing the pay table by longevity as well as by rank.

Finally, up-or-out rules are also hypothesized to induce effort by lowering the expected payoff to remaining in a lower grade (relative to advancement to a higher rank). Up-or-out rules can serve as a substitute for a direct increase in interrank pay spreads.

---

[6]Armies composed of draftees are difficult to motivate. The analysis here makes clear why. Contingent compensation cannot be used to motivate personnel who are not going to stay around long enough to collect it. Draft armies must be motivated by penalties associated with failure to perform (e.g., imprisonment and bad conduct discharges) rather than the promise of positive rewards for good performance.

Organizational Policies

We next discuss the policy implications of our analysis of individual decisionmaking. In this discussion, we highlight two key policies: (1) the sequencing of intergrade and intragrade pay, and (2) retired pay. We focus on these two tools because of their relevance to our analysis later of the military retirement system and MFERS. The relevance of a discussion of retired pay is obvious. But a discussion of the sequencing of intergrade and intragrade pay is also relevant because MFERS would need to be coupled with a pay raise so it could be implemented in a way to keep service members equally well off. Thus, a discussion of the structure of pay as a policy tool is germane.

Sequencing Intergrade and Intragrade Pay. Consider the model's implications for how pay should be sequenced by grade and longevity. Because promotions through the junior ranks occur with virtual certainty based on skill acquisition and satisfaction of time-in-grade (TIG) and time-in-service (TIS) requirements, large intergrade increases to motivate effort are not needed. Luck plays a small role in the promotion process, so there is a less blunted relationship between effort and the likelihood of promotion in the lower grades.

But beyond the junior ranks, when personnel begin to reach the middle ranks in the second five years of service, promotions start to resemble a "tournament" with winners (promotees) and losers (nonpromotees). The military's objective is to sharpen the competition and to induce the most qualified to reveal themselves in the promotion contest. Among other policies, sharper competition is induced through bigger intergrade pay spreads. Larger intergrade spreads motivate harder work in the quest for advancement and therefore discourage slumming (slacking). Larger spreads encourage the more able to remain in service and therefore help maintain the quality of the promotion pool. And by improving the talent pool and by inducing the more able to work harder, larger intergrade spreads prevent "climbing" (promotions of the less qualified).

As individuals progress toward the senior ranks, promotion rates fall. Absent any change in the structure of pay, declining promotion rates tend to discourage effort (when the probability of promotion is less than 50 percent). Clearly, interrank pay spreads need to rise with rank—that is, *be skewed*—to maintain effort. The tendency to reduce effort is accentuated by several other factors. Two mentioned previously are the rising relative importance of "luck" in promotion outcomes and the increasing homogeneity of the promotion pool. Another is that as personnel progress through the ranks the number of remaining promotions (and therefore promotion payoffs) that can be earned decreases. Skewness is required for personnel to see a continuing reward to effort.

A final factor that leads to increased skewness is the fact that the number of participants in the promotion contest declines as individuals progress through the ranks. We show elsewhere (Asch and Warner [1994a]) that the marginal value of effort is smaller in contests that have fewer participants because in small contests people can pass fewer competitors by working harder. Since the scale of the contest diminishes at higher ranks, the interrank spreads should increase to maintain effort incentives.[7]

Other factors, though, reduce the required skewness. Obviously, the more value that individuals attach to the status and other nonpecuniaries associated with higher ranks, the smaller are the additional monetary rewards needed to motivate effort in the lower ranks. These nonpecuniary factors tend to rise with grade. A second factor is the transferability of training. The less that training received in the service improves outside employment opportunities, the smaller the in-service pay increases will need to be to maintain a given level of retention. The third factor is the correlation between tastes and ability. If the correlation is positive, so that the personnel who have stronger tastes for the military are also the more able, then less skewness is required to induce the more able to stay and seek the higher ranks.[8]

---

[7]Notice, though, that in the military there is still a sizable pool of competitors for promotion to the highest ranks, so pay spreads need not be as large here to motivate effort as in the top levels of corporations, where only a handful of competitors may be vying for promotion.

[8]An oft-cited factor that reduces the optimal degree of skewness is that the production of military "output" is team-oriented. Rosen (1992, pp. 234–235) writes that "if rewards are skewed too much, competitors may take steps to make others look bad rather than making themselves look good. Lack of cooperation and reduced cohesiveness can reduce the effectiveness of the overall team. Some happy medium must be struck here." In our opinion, this argument is not particularly compelling in the military case because of the sheer numbers of individuals participating in the promo-

Intragrade pay should function like intergrade pay to motivate effort and induce the proper sorting within the organization. Intragrade pay should rise to some extent with experience to provide continuing skill acquisition and performance incentives (at least when coupled with minimum performance standards for retention). However, the intrarank longevity increases cannot be as large as the interrank increases or individuals will be encouraged to "slum." At some point intrarank longevity increases should cease altogether so that those who are revealed to be unpromotable will be induced to leave voluntarily when it is in the services' interest that they do so.[9]

Finally, personnel policies like up-or-out rules and minimum performance standards can play a positive role by: (1) increasing effort, and (2) inducing the voluntary departure of those who have low promotion chances. The extra turnover induced by up-or-out rules helps maintain promotion flows.

Retired Pay. What are the purposes of retired pay? Does retired pay have a unique role that cannot be accomplished with other forms of compensation or other policy tools? The lateral entry constraint means the military must access and train large numbers of entrants before identifying for advancement those who have the talent to perform the higher-level tasks in the organization. The military therefore wants to provide incentives for the most talented to stay and seek advancement and for others to leave after they find that they are unsuitable for the upper-level positions. That is, it must provide the proper incentives for personnel to self-sort. Salop and Salop (1976) were the first to recognize the use of "two-part" compensation schemes as a self-selection device. One such "two-part" scheme is a system of (1) active pay and (2) deferred retirement

---

tion contests, their geographic dispersion (sabotage is more likely when people work together), and the fact that performance evaluations often tend to focus on team performance. In fact, concerns about military pay spreads usually have more to do with horizontal equity than vertical equity. Some critics believe that interoccupational pay variations arising from bonuses and the like erode cooperation and *esprit de corps.* Whether interoccupational pay spreads have any effect on morale is an unresolved question. Note, though, that several foreign militaries, including the United Kingdom, have well-institutionalized systems of "skill pay," with no apparent detrimental effects.

[9]The *Report of the Seventh Quadrennial Review of Military Compensation* (1992) had in fact identified and recommended correction of a number of inconsistencies between intrarank and interrank pay.

benefits that are paid only to those who achieve a certain rank and longevity. Delayed vesting of retired pay induces self-sorting because only those who think that they can achieve the requisite rank and longevity will stay early on while others will leave. Deferred retired pay is also predicted to motivate work effort, especially when combined with minimum performance standards for retention and up-or-out rules that prevent low-ranking personnel from staying long enough to collect retirement benefits.

The question arises why retirement benefits should be part of the self-sorting mechanism. After all, why not just pay a bonus to all who reach the requisite rank and years of service? The answer has to do with retired pay's role as a separation incentive. At some point the military wants everyone, including the best personnel, to separate, even though they may still be very productive (i.e., their own productivity exceeds their pay). The longer individuals remain in the top positions, the slower will be the promotion rates for younger (and potentially equally able) personnel. Unless offset by changes in the structure of pay, reduced promotion opportunities in the junior ranks is predicted to discourage work effort in those ranks and will cause those junior personnel with the best external opportunities (i.e., the more able) to leave. Without the proper inducement, senior personnel may not want to leave voluntarily if their military pay exceeds their best private sector alternatives. Such is especially likely to be the case for those trained in military-specific skills.

Retired pay can be used to induce voluntary separation of senior personnel. For example, once personnel become vested in the immediate annuities provided by the current retirement system, they have a much reduced gain from staying and are therefore more willing to depart voluntarily.[10] The retirement system therefore induces the separations needed to control the age or experience structure of the force and to maintain promotion flows for younger personnel.

---

[10]Because their gain to staying is smaller, turnover of enlisted personnel at YOS 20 is much higher than officer turnover. Most enlisted personnel have reached their terminal grades by YOS 20 and have fewer promotions and smaller in-grade longevity raises to look forward to. Beyond the 20-year mark, officers appear to postpone their separations until they fail selection to the next rank.

There is, of course, no reason why the separations required to maintain personnel flows could not be accomplished with other policy tools, like up-or-out rules. In fact, during the drawdown period, mandatory separations increased substantially with the reduction of high-year-of-tenure points. However, excessive reliance on involuntary separation to control the experience structure of the force can be bad for morale, and affect recruiting, retention, and work effort. These adverse effects might require the payment of a "regret premium" to compensate for the prospect of involuntary separation. In addition, personnel faced with the prospect of involuntary separation are likely to engage in activities aimed at getting the policy relaxed (e.g., complaining to personnel managers and writing to congressmen about the "unfairness" of the policy). Should their complaints prove successful, the services would be compelled to modify their forces in unproductive ways. After Milgrom (1988), we call these extra financial costs and productivity effects the *organizational influence costs* of mandatory separation. The organizational influence costs of the drawdown are apparent today, with discontent in the mid-ranks over the likelihood of mandatory separation. Separation pay is the "elixir" that eases termination from service and weakens potential criticisms about the capriciousness or arbitrariness of policy.

As mentioned in the introduction, critics of the current retirement system have charged that efficiency would be increased if the military shifted compensation away from retired pay and toward active duty pay, since members heavily discount future retired pay. However, such a policy would necessitate heavier reliance on involuntary separation to control the experience distribution of the force. Pressure would develop on the services to relax their policies and permit older personnel to stay until full retirement and superannuated forces might result.[11] The adverse productivity effects of a much older force or the regret premium that might be required to maintain the current (younger) experience distribution, while hard to

---

[11]Data from the reserves provide evidence that, in the absence of separation incentives, personnel would want to remain for much longer careers. Although vested after 20 creditable years of reserve service, reservists do not begin to receive benefits until age 60. Retention of reservists with 20 or more years of service is much higher than in the active force. In fact, there is some concern about superannuation in the reserve forces.

calculate, could be substantial. While clearly expensive, a system that provides voluntary separation incentives is likely to be cheaper.

The other purposes of retired pay are, of course, not unique. Motivating effort, improving retention, and inducing personnel to properly self-sort within the organization could be accomplished through an appropriately structured active duty pay table and through other personnel policies. The distinctive (if not unique) purpose for military retired pay is to induce voluntary separations at the appropriate points, thereby minimizing the influence costs that accompany involuntary separation.

EMPIRICAL MODEL OVERVIEW

To implement our theoretical model empirically we could have used two alternative approaches. One is to estimate the parameters of our model empirically using panel data on observed individual retention decisions and effort decisions (how much effort to expend) over the course of an individual's career and then use the estimated model to forecast the effects of different policies. Such an approach is infeasible for two reasons. First, other than work by Gotz and McCall (1984) and more recent work by Daula and Moffitt (1995), attempts to estimate just the retention portion of the model have not borne much fruit. Second, except for some spotty information on military personnel performance, data on effort decisions do not exist.

We therefore took the second and more parsimonious approach— computer simulation of our theoretical model. To build this microsimulation model, we needed three types of parameter values: those relating to individual retention decisions, to individual effort decisions, and to the relationship between ability and compensation. For the retention-related parameters, in our model the individual's stay/leave decision is basically characterized by three parameters: the mean and standard deviation of the initial taste distribution and the standard deviation of the distribution of random shocks that each individual faces in each grade and year of service. (See Asch and Warner [1994a] for a more formal description.) We experimented with alternative values of these parameters until the model replicated the historically observed aggregate retention patterns. To model personnel effort decisions empirically, we made assumptions about the relationships between effort supply and promotion and about the cost to individuals of supplying effort. We then conducted sensitivity analyses to determine whether the model's results were sensitive to changes in assumed parametric values. Finally, to implement empirically the ability sorting aspect of the model, we used data on military personnel aptitude scores, which are considered to be correlates of ability, as well as estimates made by previous studies of the effect of these scores on promotion probabilities. The discussion below provides more detail about how we constructed our empirical model, which we use to generate the results in Chapters Four and Five.

Retention and Force Structure

We calibrated the model using data from the Army enlisted force. To better estimate any cost savings associated with moving the current retirement system to MFERS, we also calibrated the model using data from the Navy enlisted force and from the Air Force enlisted force. In Chapter Four, we give results only for the Army because the results are qualitatively the same for the other services. However, we include the cost results from our analysis of the Navy and Air Force as well as the Army in our estimates of the cost savings associated with moving to MFERS.

The steady-state grade-by-YOS distribution of a given force will depend on many factors. The three crucial factors are the lengths of the initial enlistment and reenlistment contracts, promotion rates and timing, and retention rates. To implement the model empirically, we had to make simplifying assumptions about enlistment contracts. In reality, the almost infinite variety of enlistment and reenlistment contract lengths would be extremely difficult to model without individual-level data. Enlistees in the Army join for periods of two to four years, whereas Navy enlistees join for periods of three to six years. Once initial enlistments are completed, enlisted personnel can either extend their current enlistment contract for up to two years or reenlist for periods of three to six years. We simplify the model considerably by assuming that enlisted personnel initially enter for 4 years and then reenlist thereafter for four-year periods. However, we assume that once personnel reach YOS 20 and are eligible to retire, they make annual retention decisions thereafter. This assumption seems to be supported by the data—the continuation rates of those not at their estimated time of separation (ETS) are much lower after YOS 20 than before, indicating less-rigid enforcement of enlistment contracts and more frequent retention decisionmaking beyond YOS 20.

We chose fiscal years 1987–1989 as representative for data on promotion rates and force structure. Promotion rates began to decline after 1989 as a result of the drawdown and therefore may not be representative of steady-state promotion opportunities.[12] We used data from the Defense Manpower Data Center (DMDC) to compute promotion probabilities for enlisted personnel in each service. DMDC makes available data by fiscal year on end-strengths, promotions, and losses by grade and YOS. The promotion rate from a given grade-YOS cell was calculated as a proportion of personnel in the given grade-YOS cell at the end of each fiscal year that both stayed and was promoted during the next fiscal year. We then calculated the three-year (FY 1987–1989) average of these rates.

Finally, to calibrate the model, we built steady-state forces that mimic as closely as possible the force structure and retention patterns that prevailed in FY 1987–1989. The calibration takes place as follows. Consider personnel entering service during a given fiscal year. Between the time of entry and the end of the fiscal year both promotions and attrition occur. We used actual FY 1987–1989 data on enlisted personnel in each service to distribute new entrants by pay grade and YOS at the end of YOS 1 and to specify the YOS 1 loss rate. We then compute flows into the different grades in YOS 2 based on FY 1987–1989 promotion rates and the FY 1987–1989 average of enlisted *non-ETS* continuation rates for YOS 2. These flows are then adjusted to account for prior-service gains based on an average of FY 1987–1989 prior-service gain rates into YOS 2. We repeat the process for YOS 3.

Choice behavior begins to occur in YOS 4. Choice is based on the expected gain to staying. Conceptually, each member of the cohort that survives to YOS 4 has a gain to staying (or cost of leaving) that is based on: (1) the military pay table, the retirement system, and the

---

[12]The exception is the Air Force. Because the Air Force started drawing down its forces in 1989, we chose fiscal years 1987 and 1988 as a representative data period for the Air Force.

civilian pay stream that he or she faces; (2) future promotion prob-abilities and service high-year-of-tenure (HYT) policies; (3) the member's taste for service; (4) the service member's ability; and (5) the distribution of the random factor in retention decisions. As described in the discussion of the theoretical model, the gain to staying is a probabilistic weighting of the payoffs to staying to the various future YOS points and then separating, where the probability weights depend on the strength of tastes for service and therefore vary according to a taste factor and an ability factor. The cohort retention rate is derived as a weighted average of the probabilities of staying for different values of these factors. An efficient method for performing these calculations is described in Black, Moffitt, and Warner (1990).

The proportion of the YOS 4 cohort that stays (in a probabilistic sense) is then "aged" by YOS and grade over the next four years based on FY 1987–1989 promotion rates by grade and YOS and FY 1987–1989 non-ETS continuation rates. The fraction that survives to (each grade in) YOS 8 is then allowed to make another retention decision, which is again based on the factors identified above. The process repeats itself over the next four-year interval, and so forth.

Finally, the continuation rate in a given grade-YOS cell is set to zero if the YOS is equal to or greater than the grade's HYT or up-or-out point. To make the model fit the observed FY 1987–1989 force better, in some cases the HYT is relaxed a year or two because significant numbers of personnel are observed who have YOS above the nominal HYT. For example, although the Army's nominal HYT for E-8s is 24, in the FY 1987–1989 era there were significant numbers of E-8s in YOS 25 and YOS 26. Therefore, we set the E-8 HYT to be 26.

As mentioned above, the retention pattern and the resulting force structure predicted by the model are controlled by varying the three model parameters—the mean of the initial taste distribution (MUT hereafter), the standard deviation of this distribution (SDT hereafter), and the standard deviation of the random disturbance distribution (SDE hereafter). For example, increasing MUT raises retention at all YOS points (although early retention is most affected). Raising the variation in tastes, SDT, may increase or decrease retention, depending upon the levels of military and civilian pay. The YOS pattern of retention depends on the importance of random factors in the retention process relative to tastes. Random factors are less important the smaller SDE is. The smaller SDE is, the more retention tends to rise with YOS beyond the initial retention decision. In fact, if SDE were zero, then retention rates would jump to unity after the initial retention decision (as long as the gain to staying rises with YOS). That voluntary retention rates do not increase so sharply indicates that random factors are important.[13]

The first panel of Table 2 shows the actual grade-by-YOS of Army enlisted personnel for the FY 1987–1989 period. The distribution is virtually the same as the FY 1990 distribution. Based on an average of FY 1987–1989 continuation rates, the table shows what fraction of an entry cohort would survive to various years of service. About 34 percent would survive to YOS 5; 12 percent would survive to YOS 20 and become retirement-eligible. If the continuation rates were steady-state, the Army would get 5.31 man-years per accession on average. The average enlisted strength during this period was 647,187, and the Army would require 121,785 accessions per year to sustain this size force based on the FY 1987–1989 continuation rates.

An unsettled question is the rate at which personnel discount future dollars. Some previous research (Gilman [1976], Black [1983], Lawrence [1991]) suggests that personnel discount future dollars at fairly high rates. In their estimation of the dynamic retention model, Daula and Moffitt (1995) claim an econometric estimate of 9.9 percent. Using data on the drawdown choice of a lump sum versus an annuity separation benefit, Warner and Pleeter (1995) obtain an estimate of as high as 20 percent. We calibrated the model at a rate of 10 percent for all personnel.

The second panel of Table 2 shows the model parameters that yield simulated Army retention patterns and an Army force structure that was as close to the observed FY 1987–1989 force as we could get. Although the model fit is not exact, it is close: The force has virtually the same experience mix, the same survival to YOS 20, and a roughly similar grade distribution. Man-years per accession are slightly

---

[13]The Army's FY 1987–1989 average ETS retention rate at YOS 4 was 35 percent. At YOS 8 it was 64 percent and at YOS 12 it was 80 percent.

Table 2

Model Fits for Army Enlisted Personnel

I. Enlisted Force Based on Actual FY 1987–1989 Army Data						
	Grade-by-YOS Distribution				Survival to	
YOS	E-1–E-3	E-4–E-6	E-7–E-9	Total	Start of	
YOS 1–4	28.0	21.7	.0	49.7	YOS 5	.338
YOS 5–10	.7	25.7	.1	26.5	YOS 10	.189
YOS 11–20	.0	13.0	8.5	21.5	YOS 20	.120
YOS 21–30	.0	.1	2.2	2.3	YOS 30	.005
Total	28.7	60.5	10.8			

NOTES: Man-years per accession = 5.31, accessions based on force of 647,187 = 121,785.

II. Force Based on the Assumptions: Personal discount rate = 10 percent; MUT = 0; SDT = 3000; SDE = 40,000

	Grade-by-YOS Distribution				Survival to	
YOS	E-1–E-3	E-4–E-6	E-7– E-9	Total	Start of	
YOS 1–4	31.3	20.7	.0	52.0	YOS 5	.307
YOS 5–10	.0	24.8	.3	25.1	YOS 10	.164
YOS 11–20	.0	14.0	7.1	21.1	YOS 20	.107
YOS 21–30	.0	.0	1.8	1.8	YOS 30	.001
Total	31.3	59.5	9.2			

NOTES: Man-years per accession = 5.35, accessions based on force of 647,187 = 120,925.

higher in our simulated force and required accessions are slightly lower than in the actual force. This result arises from our assumption that all entrants enlist for four years. A significant proportion of Army entrants enlist for two or three years, which lowers the Army's actual man-years per accession. Nevertheless, the point is not to perfectly predict the actual force, but to build a hypothetical force with characteristics as close as possible to the observed one with our simplifying assumptions and then study how that force would react to changes in compensation and personnel policy.

We find that we can also closely replicate the Navy force size and structure. For the Air Force, our calibration is less exact. The model overpredicts first-term retention and slightly underpredicts the flow of members who stay for 20 years of service. But adjusting the retention parameters to reduce first-term retention reduces the flow to 20-year retirement too much. The basic problem is that our assumption that retention decisions are made only every four years is too restrictive for the Air Force. Examination of non-ETS continuation rates suggests that Air Force enlisted members make retention decisions at times other than their formal ETS. Since no set of parameter values will exactly replicate the Air Force enlisted force, we chose a set of assumptions that would simulate the flow to the 20-year point relatively accurately. This set of assumptions should permit us to estimate fairly well the cost savings associated with moving the Air Force to MFERS.

A key test of the model's plausibility is whether its predictions of the response to changes in compensation are consistent with available empirical evidence. To find out, we simulated the effects of (1) a one-multiple increase in the Selective Reenlistment Bonus (SRB) available at YOS 4, (2) a one-multiple increase at YOS 8, and (3) a 10-percent across-the-board increase in basic pay. The predictions generated by the model are within the range of estimates provided by econometric evidence. Evaluated on grounds of plausibility of the responsiveness of retention to changes in pay, the model seems well calibrated.

Computing Ability and Effort Supply

In addition to estimating the force structure implications of alternative compensation structures, the simulation model also estimates the implications for ability sorting and the average amount of effort supplied by the force. To incorporate the role of ability, we first posit a standard normal probability distribution of ability among the entry cohort. We then allow different ability types (captured by deviations from the mean ability level) to affect earnings in alternative employment (i.e., civilian earnings) and to affect the probability of promotion in each grade and YOS. We proxied these effects by using previous estimates of the relationships between Armed Forces Qualification Test (AFQT) scores and civilian earnings and between AFQT scores and promotion probabilities.[14] Although AFQT score is not a direct measure of ability, it is thought to be a strong correlate of it.

---

[14]These previous estimates are obtained from Smith, Sylwester, and Villa (1991).

In calibrating the model, we had to make an assumption about the correlation between individuals' tastes for service and their ability. We calibrate the model assuming no correlation between tastes and ability.

Of particular interest from a policy standpoint is how compensation and personnel policy affect how well the organization is able to provide an incentive for the most able to stay and seek advancement. To measure the "ability sorting" effects of alternative retirement policies, we therefore compute the average ability of personnel by grade. Since the units in which ability is measured are set arbitrarily, the average E-1 ability level is set to zero. To measure ability sorting, we simply measure average E-9 ability—if average E-1 ability is zero, average E-9 ability tells us the degree to which the compensation and personnel systems induce high-ability individuals to stay and seek advancement to the upper grades.[15] It should also be noted that since the units that ability is measured in are arbitrary, the *changes* in ability and ability sorting as a result of *changes* in policy will be of primary interest rather than the absolute levels of ability under each policy, per se.

Incorporating effort supply into the model is more complicated because, like the retention decision, the optimal effort supply decision for each individual is made in each grade and year of service and is both a forward-looking and backward-looking decision process. Furthermore, the decision will differ for individuals of different taste and ability types. To incorporate these factors, we first defined "individuals" in terms of standard deviations from the mean of the taste distribution and standard deviations from the mean of the ability distribution. We then calculated each "individual's" optimal effort level in each grade and YOS interactively using Newton's method.[16]

As discussed earlier in this chapter, the optimal effort is given at the point where the marginal benefit of effort equals the marginal cost of

---

[15]The choice of E-9 is arbitrary. We could also have measured ability sorting by measuring average E-7 or E-8 ability. Our results are qualitatively the same when we choose these grades instead.

[16]For a generic description of how to use this method to numerically solve derivatives, see Press et al. (1992), p. 355.

supplying it. Two factors most affect the marginal benefit: (1) the effect of effort on the probability of promotion, and (2) the return to being promoted (including the increment in basic pay and in status and rank in the current period and in future periods as a result of the promotion). The second factor is given by policy in our model. Thus, calibrating the model's effort parameters required making assumptions about the effect of effort on the probability of promotion and about the marginal cost of effort.

Given our general lack of knowledge about what values these parameters should take, these assumptions will necessarily be arbitrary. Indeed, one of the reasons for using a simulation rather than an estimation approach is the lack of data on effort. Although the effort-related parameters are somewhat arbitrary, recall that our focus is on how optimal effort changes when policy changes and not on the absolute level of effort supplied. Thus, we want to set the parameters so that the results are not strongly affected by changes in their assumed values.

Consider our specification of the marginal cost of effort. Marginal cost is assumed to be linear in effort (i.e., marginal cost equals $10e_{it}$ where e_{it} is effort in grade i in period t).[17] Raising and lowering the linear term by a factor of 10 (from 10 to 100 and 1, respectively) had no significant effect on the force structure and cost results shown in the next chapter.

We also had to make some assumptions about the effect of effort on promotion probabilities. In our theoretical model, the military evaluates the individuals seeking advancement and then selects some fraction for promotion. Although evaluation scores are subject to random factors, individuals can increase their scores in these contests and thus their probability of promotion by either being more able or by supplying more effort. The individual's probability of promotion also depends on the ability and effort supply of all the other individuals vying for promotion.

Incorporating such contests into our empirical model at each grade and YOS for each individual's ability and taste would add many lay-

---

[17]More specifically, in the enlisted model we assume that the disutility (or cost) of effort is given as $5(e_{it})^2$, so that the marginal cost of effort equals $10e_{it}$.

ers of complexity into our model. It would involve making specific assumptions about the military's evaluation process and how effort and ability interact to affect an individual's evaluation and thus one's promotion chances. To minimize the number of assumptions we had to make, we first assumed that individual effort decisions have no effect on the Army's aggregate promotion rate into each grade at each year of service. Given the large numbers of individuals who are competing for promotion at any given point, this assumption seems reasonable. However, we also assume that individuals view their own effort as having a positive effect on their individual chances of promotion and thus their marginal benefit of effort. After some experimentation, we set the effect of effort on the probability of promotion (denoted Beta_E) equal to .01. Increasing Beta_E to .1 increases the average optimal effort in the force but has little effect on retention patterns. Similarly, reducing this parameter by a factor of 10 (to .001) reduces average optimal effort but has little force structure or cost impact. The average ability level of the force also changes little as well.

We also note that although we can incorporate into our theoretical model the nonpecuniary rewards to increased effort such as better assignments, it is more difficult to incorporate them into our empirical model. We therefore ignore them. In addition, although our model can predict the effects of various policy variables on individual effort, it does not predict the implications of changes in individual effort for unit performance and military output. Presumably, higher individual effort will translate into better unit performance and increased military output, but our model does not specify how this occurs.

Cost Analysis

Once the model builds a steady-state force, it provides costs for that force. The two costs we focus on in the analyses below are the annual basic payroll cost and annual accrual cost of the retirement system. Until recently the DoD Actuary used a 2 percent real interest rate in its calculations, so we use a 2 percent real rate in our main analysis. The sensitivity of the cost results to changes in the assumed real interest rate is discussed in Chapter Four.

Estimating the MFERS Retirement Annuity

To calculate the expected retirement annuity a member would receive under MFERS, several assumptions had to made. First, we had to make assumptions about members' contribution rates under the thrift savings plan. We assumed that the contribution rates for MFERS were the same as for FERS, holding age constant. These rates are shown in Table 3. The contribution rate for a given age interval equals the average fraction of basic pay an individual in that age range contributes to FERS times the fraction of those in the age range that contribute. Thus, the contribution rate is 3.25 percent for those between ages 20 and 29 and 5.75 percent for those between ages 50 and 59.

Of course, experience under MFERS may differ from the experience under FERS. A more complete analysis would model the member's choice to participate in the thrift savings plan and the amount he or she would contribute. We ignore this facet of individual decision-making in our model. Instead, we conduct sensitivity analyses to test how sensitive our qualitative results are to different assumptions about member contribution rates. We find that the qualitative results are unchanged.

Under MFERS, members have several withdrawal options. First, they can choose to retire early and take reduced benefits under the basic plan. Second, under the thrift savings plan they can opt to take their benefits as an annuity, roll over their benefits into an IRA which they can begin withdrawing at age 59, or take a lump sum payment but face a penalty. We assume that members choose the option that yields the highest expected discounted present value.

Table 3

Assumed MFERS Thrift Savings Plan Contribution Rates

Age	Rate (percent)
20–29	3.25
30–39	4.0
40–49	4.5
50–59	5.75

To calculate a member's accumulated contributions over his or her career, we assume that these contributions accumulate at the government real rate of interest (which we assume equals 2 percent). We assume an extremely conservative real return to thrift savings plan investments since 2 percent is the approximate return on Treasury bills. Long-term government bonds have somewhat higher real returns, and the real return in the stock market over the past 20 years has averaged about 9 percent. In calculating the accumulated contributions, we also assume that individuals progress through the military pay table at the average rate of promotion for each grade and year of service.

To incorporate the COLAs offered under the basic benefit plan, we assume an annual inflation rate of 3 percent. Finally, we assume that members who take the IRA withdrawal option under the thrift savings plan will have an IRA accumulation at a rate of interest equal to the government interest rate plus 2 percent (or 4 percent).

Modeling the Transition to MFERS

The above discussion addresses how we developed our steady-state empirical model. This model predicts behavior when all personnel are under a given compensation system. We also developed a model that predicts results for the transition to the new steady state to produce the results presented in Chapter Five. As discussed there, in considering the transition to the steady state, we consider two transition strategies. In the first case (the grandfathering case), new entrants are enrolled into the new compensation system but existing members are grandfathered into the current system. In the second case (the convert case), new entrants are enrolled automatically in the new system, but existing members are permitted to convert to the new system. For simplicity in constructing the transition model, we consider only retention and cost effects and ignore effort supply and ability sorting effects.[18]

---

[18]The model can be extended to consider effort supply and ability sorting effects. Although the model currently does not produce results relating to ability sorting, it accounts for the ability distribution for each cohort at each calendar year in computing the projected retention patterns for each year.

In the transition model, we project behavior by calendar year. For example, in the grandfathering case, the model projects behavior year after calendar year, as each entering cohort marches through its careers. Since we assume a maximum military career length of 30 years, it takes 30 years for all personnel to be under the new system. Because the transition model predicts retention outcomes in each year, it can be thought of as an inventory projection model.

To predict retention patterns in each year, we use the following general approach. First, we calculate the probability an individual in a given grade and year of service and with a given taste for military service stays in the service over the rest of his or her career under the current military compensation system and under MFERS. This probability is calculated using the methods used in the steady-state model. Second, we use these probabilities to predict the force size and structure for each future calendar year, accounting for the distribution of taste for military service. We assume that accessions vary each calendar year to ensure that the force size is constant from year to year. Third, we also use these probabilities to compute DoD's basic pay costs and annual retirement costs. Under MFERS, the latter equals DoD's contributions to the thrift savings plan as well as the retirement accrual charge for the basic plan.

To predict which members convert and which do not under case 2, we assume that a member with a given taste, grade, and year of service converts if the gain to staying under MFERS is greater than under the current plan (which, for simplicity, we assume is REDUX).

As a test of the consistency of the transition model with the steady-state model, we compared the results predicted for years 30 and beyond, when all members would be under the new system, with those produced by the steady-state model. Since accessions are allowed to vary from year to year in transition to the steady state, the new steady state may not be reached until some point well beyond year 30. Nonetheless, we find that the transition model's predicted retention rates and costs for year 30 are extremely close to the steady-state predictions (within .5 percent). Thus, we have confidence that the transition and steady-state models are mutually consistent.

Chapter Four

STEADY-STATE RESULTS

We next use our calibrated model to analyze the effects of MFERS relative to the current system in the steady state. We begin by predicting the steady-state force structure and productivity consequences of REDUX, which is the current system for those entering service after 1986. We then analyze MFERS for military personnel in the case of no pay raise. We call this basic MFERS.

Following our analysis of basic MFERS, we consider MFERS with two different types of pay raises. The first type is an across-the-board pay raise—all members get identical percentage raises. This is the type of raise generally given military members each year to adjust for cost-of-living changes and other pay trends in the civilian sector. The second type is a skewed pay raise. In this case, those in the upper grades would get larger percentage raises. As discussed in Chapter Three, there are several distinct advantages associated with skewing the pay structure. We find that not only can MFERS with a skewed pay raise produce the same general steady-state force structure as REDUX or MFERS with an across-the-board pay raise, but it raises the measure of productivity while lowering costs.

THE CURRENT MILITARY RETIREMENT SYSTEM: REDUX

There are three military retirement systems now in effect. In brief, pre-1980 entrants who complete 20+ YOS receive a lifetime, inflation-protected annuity according to the formula .025*YOS*final basic pay. FY 1981–1986 entrants who complete 20+ YOS sometime after the year 2000 will receive a lifetime, inflation-protected annuity according to the formula .025*YOS*high-3 years' average basic pay.

Post-1986 entrants who serve less than 30 years of service will receive a reduced annuity until age 62 but the same annuity as FY 1981–1986 entrants beginning at age 62. These annuities, however, will not be fully inflation-protected; rather they will be allowed to erode in real value at the rate of 1 percent per year. Finally, personnel who reach a HYT point in the YOS 11–19 range (e.g., O-3s at YOS 11) receive involuntary separation pay according to the formula .1*YOS*final basic pay.

The model was calibrated using the pre-1980 military retirement system since this is the system that applies to most of those in 1987–1989, the years we chose as representative for calibrating the model's retention parameters.[1] Table 4 shows the predicted Army force structure under REDUX. Relative to the system for pre-1980 personnel shown in Table 2, REDUX is predicted to reduce the probability that an enlisted entrant will stay for 20 years from .107 to .086, a decline of about 20 percent. Accessions required to maintain a constant force level rise by about 5 percent. We estimate that the deleterious retention effect of REDUX could be offset by an active duty pay raise of 3 percent.[2] Another effect of REDUX is that it signifi-

Table 4

**Predicted Effects of Post-1986 Retirement System
(REDUX)**

YOS	Grade-by-YOS Distribution			Total	Survival to Start of	
	E-1–E-3	E-4–E-6	E-7–E-9			
1–4	33.1	21.6	.0	54.7	YOS 5	.294
5–10	.0	24.4	.3	24.7	YOS 10	.145
11–20	.0	12.2	6.4	18.6	YOS 20	.086
21–30	.0	.0	2.0	2.0	YOS 30	.002
Total	33.1	58.2	8.7			

NOTES: Man-years per accession = 5.06, accessions based on force of 647,187 = 127,883, average effort indicator = 3.35, average E-9 ability = .089.

---

[1]In reality, most personnel are covered by either the pre-1980 or the FY 1981–1986 plans. However, as shown in Asch and Warner (1994b), there are no substantial behavioral differences of personnel under these two systems.

[2]The amount of the pay raise will depend on the personal discount rate. Here we assume a discount rate of 10 percent.

cantly increases post-YOS 20 retention. These effects were to be expected given the increase in the retirement multiplier and less-than-full inflation indexing.

The estimated effects of REDUX on the average optimal effort of the force and ability sorting are shown in the Notes to Table 4, and the cost of REDUX is shown in Table 5. These figures together with the force structure figures form the base case against which we compare MFERS below.

MFERS WITHOUT A PAY RAISE

MFERS without a pay raise represents a lower-value compensation package than the current military compensation system because, for those with 20 or more years of service, the value of retired pay under MFERS is significantly less than under REDUX. Table 6 shows the discounted present value of the benefits that separatees from various ranks and years of service would receive under the two systems. The amounts are based on the January 1992 basic pay table. MFERS would pay an annuity to those who separated as early as after three years whereas the current system would pay nothing. On the other hand, MFERS is estimated to reduce the real value of the present value of retired pay by about 50 to 60 percent for an individual who is at or beyond 20 years of service.

For those with less than 20 years of service, MFERS is more generous since MFERS vests earlier than the current system. However, overall, MFERS is a less-generous system. The benefits for those with less than 20 years of service tend to be small and, in terms of retention effects, the larger value of the present value of retired pay for those

Table 5

Steady-State Costs of REDUX

Item	Annual Payroll Cost (billions)
Basic pay	$9.34
Retirement accrual	$1.95
Total	$11.29

with less than 20 years under MFERS does not offset the smaller value for those with more than 20 years of service.

Because MFERS is a less-generous system, retention is significantly reduced, as may be seen by comparing the predicted retention rates and force structure under MFERS without a pay raise in Table 7 with those under REDUX. Survival to YOS 10 falls by 30 percent and survival to YOS 20 falls by 65 percent. Consequently, the fraction of the force in the more junior grades and earlier YOS rises by about 10 percent, and man-years per accession falls from 5.06 years to 4.24, over three-quarters of a year. Average effort is also predicted to be lower under MFERS and less ability sorting is predicted to take place because the expected value of retired pay is lower under MFERS so that deferred compensation is less. The main conclusion is that unless planners desire a more junior force under MFERS, moving to MFERS

Table 6

Discounted Present Value of Retired Pay Under REDUX and MFERS with No Pay Raise

Grade/YOS	REDUX	MFERS
E-4/5	$0	$4,510
E-5/10	$0	$11,074
E-6/15	$0	$21,228
E-7/20	$88,242	$35,274
E-8/25	$149,011	$60,763
E-9/30	$237,019	$113,274

Table 7

Predicted Effects of MFERS without a Pay Raise

YOS	Grade-by-YOS Distribution			Total	Survival to Start of	
	E-1–E-3	E-4–E-6	E-7–E-9			
1–4	.395	.251	.000	.646	YOS 5	.265
5–10	.000	.240	.002	.242	YOS 10	.100
11–20	.000	.068	.034	.102	YOS 20	.030
21–30	.000	.000	.009	.009	YOS 30	.001
Total	.395	.559	.045			

NOTES: Man-years per accession = 4.24, average effort indicator = 2.54, average E-9 ability = –.080.

must be accompanied by a pay raise to maintain the force size and structure and to motivate effort supply and maintain quality.

MFERS WITH AN ACROSS-THE-BOARD PAY RAISE

How big a pay raise is needed? As discussed in Chapter Two, we wished to find a pay raise that was sufficient to offset the contributions that members must make under MFERS and the tax implications. Such a pay increase—equal to about 7 percent—would keep a member's paycheck constant under MFERS relative to REDUX. However, a 7 percent pay raise is not sufficient to offset the fact that the expected value of retired pay is lower under MFERS. Retention is lower under MFERS with a 7 percent pay raise than under REDUX.[3] For example, the model predicts that man-years per accession would only be 4.65 years instead of 5.06 under REDUX. It also predicts that the probability of an individual surviving to YOS 20 would fall by 49 percent, implying that the fraction of the force with YOS between 10 and 20 falls by 55 percent. Thus, to the extent that the services wish to maintain the same force size and structure as under REDUX, another policy tool is needed to address these retention problems.

As discussed in Chapter Two, these retention problems could be addressed with a system of retention or reenlistment bonuses. However, because our model cannot easily accommodate reenlistment bonuses or other occupation-specific pays, we assume that these retention problems are addressed by a pay raise that is larger than 7 percent. Conceptually, a pay raise and a system of retention bonuses are similar. They differ to the extent that bonuses can be targeted to distinct populations, and therefore are less costly than a pay raise that is unilaterally applied to all members. Bonuses also introduce more uncertainty in a member's compensation as the services turn them on and off. However, reenlistment bonuses and pay are similar conceptually in that both create a front-loaded compensation system—one that places a greater fraction of compensation in the form of active pay and less in the form of retired or separation pay. Since one of the key aspects of MFERS is that it would make the military compensation system more front-loaded, and indeed this

---

[3]Of course, if a smaller, more junior force is desired, a 7 percent pay increase might be sufficient.

front-loading is the main source of the cost savings associated with MFERS, it seems reasonable to analyze the case of MFERS plus a more-generous pay raise rather than MFERS plus a 7 percent pay increase plus a system of retention bonuses.

It turns out that a pay raise of 13 percent would be sufficient to maintain retention incentives and thus the force size and structure. Under MFERS with a 13 percent pay raise, man-years per accession are 5.06 years, equal to that under REDUX.[4] However, although a 13 percent across-the-board pay raise can achieve the same force size and structure as REDUX, our model predicts that effort incentives and ability sorting incentives (and thus productivity incentives) are lower.[5] Average effort would fall from 3.32 under REDUX to 3.23 under MFERS with a 13 percent pay raise and average E-9 ability would fall from .089 to .07, a 21 percent fall. Although the change in effort is not large on average, it is more dramatic for those in the senior grades where presumably a reduction in effort would be more detrimental to unit performance and military output. For example, average effort among those who are an E-6 would fall by 15 percent under MFERS with a 13 percent pay raise. On the other hand, the change in average effort among those in the more junior grades is much smaller, and in fact is predicted to increase in some cases under MFERS relative to REDUX.

The reason incentives are generally more blunted under MFERS with the across-the-board pay raise is that under the current system the main source of deferred compensation (and thus effort and ability sorting incentives) is the retirement system and its vesting in a sizable and immediate annuity at the 20-year point. MFERS eliminates the 20-year vesting and reduces the expected value of the retirement benefit. It therefore reduces the main source of deferred compensation for military personnel. The fact that the percentage pay raise is across the board and the same for each grade means that the pay raise does not offset this reduction. Because MFERS with an across-the-board pay raise generates the same force as REDUX but

---

[4]For the Air Force and the Navy, the necessary across-the-board pay raise would be 15 percent.

[5]Similarly, a system of retention bonuses that were the same for all targeted groups would likely reduce productivity incentives.

reduces our predicted productivity measures, we conclude that it does not represent an improvement over REDUX.

MFERS WITH A SKEWED PAY RAISE

For MFERS to be an improvement over REDUX, it must be coupled with a skewed pay raise, that is, higher raises for those in higher grades. Or, if retention bonuses were included in the system, MFERS would need to be coupled with a skewed pay raise and a set of reenlistment bonuses or an across-the-board 7 percent pay raise and skewed set of reenlistment bonuses. Table 8 shows the set of raises by grade that would, coupled with MFERS, generate the same general predicted retention patterns and force structure as REDUX, as shown in Table 9. Man-years per accession are 5.06 under REDUX and 5.01 under MFERS with a skewed pay raise, and the fraction of the force that is in each YOS and grade grouping are quite similar across the two systems.

MFERS with a skewed raise has two distinct advantages over REDUX, while producing the same force size and structure. It would increase productivity and reduce costs. Put differently, our measures indicate that MFERS with a skewed pay raise would be more efficient than the current system.

Table 10 shows the average effort and average E-9 ability indicators under REDUX, MFERS with a 13 percent pay raise, and MFERS with a skewed pay raise. While all three are predicted to produce the same

Table 8

**Percentage Basic Pay Increase Needed
to Maintain Constant Quality Force
Under MFERS**

Grade	Percentage Raise
E-4	2
E-5	8
E-6	14
E-7	20
E-8	26
E-9	32

Table 9

Predicted Force Structure Under REDUX and MFERS with a Skewed Pay Raise

Percent of Force in:	REDUX	MFERS + Skewed Raise
YOS 1–4	54.8	55.8
YOS 5–10	24.7	26.2
YOS 11–20	18.6	15.9
YOS 21–30	2.0	2.2
Man-years per accession	5.06	5.01

Table 10

Predicted Productivity Measures Under Alternative Systems

Productivity Measure	REDUX	MFERS with 13% Pay Raise	MFERS + Skewed Pay Raise
Average effort	3.32	3.23	3.87
Average E-9 ability	.089	.07	.166

general retention patterns, MFERS with a skewed pay raise is predicted to increase average effort by 17 percent and to increase ability sorting (E-9 average ability) by 87 percent. In contrast, as discussed above, MFERS with an across-the-board pay raise is predicted to reduce both productivity measures.

Cost figures are presented in Table 11. The retirement accrual costs are computed under the assumption of a 2 percent real government discount rate. Basic pay costs for Army enlisted personnel are estimated to be $9.34 billion under REDUX whereas total costs (basic pay plus the annual retirement accrual charge) are estimated to be $11.29 billion. MFERS with a skewed pay raise would reduce total costs by 5 percent because the retirement costs would be substantially lower. As a side note, MFERS with a skewed pay raise is also less expensive than MFERS with a 13 percent across-the-board pay raise because in the former case the pay raise is targeted to the service members whose retention patterns would be most affected by the switch to MFERS.

Table 11

Predicted Costs Under Alternative Systems
(Assumes 2 Percent Real Government Interest Rate)
($ billion)

Cost	REDUX	MFERS with 13% Pay Raise	MFERS + Skewed Pay Raise
Basic pay	9.34	10.39	9.97
Retirement accrual	1.95	0.70	0.71
Total	11.29	11.09	10.68

The figures in Table 10 are for the Army enlisted force. Estimating DoD costs requires a simulation model for each service and for officers and enlisted personnel. We created such simulation models for the Navy enlisted force and the Air Force enlisted force. Using these models, we estimate that total costs for the Navy enlisted force would fall by 4.6 percent (from $9.37 billion to $8.94 billion) by a move from REDUX to MFERS with a skewed pay raise. We estimate that total costs for the Air Force enlisted force would fall by 8.3 percent (from $9.37 billion to $8.59 billion). The cost savings is estimated to be somewhat larger for the Air Force because retention to the 20-year point is higher for this service, so the cost savings associated with front-loading compensation is somewhat greater.

We therefore estimate an annual steady-state cost savings of $1.8 billion for the Army, Navy, and Air Force enlisted forces together. This figure is based on the assumption that all enlisted personnel were initially covered in the steady state by REDUX. Of course, in reality, some current force personnel are covered by the pre-1980 retirement plan, some are covered by the 1981–1986 plan, and some are covered by REDUX. If we take the actual or projected total cost figures (e.g., basic pay costs plus retirement accrual costs) used in the President's budget for 1992 through 1997, and apply the cost savings percentages implied by our model for each force (i.e., 5.4 percent for the Army enlisted force, 4.6 percent for the Navy enlisted force, and 8.3 percent for the Air Force enlisted force), we estimate a cost savings of $2.08 billion in 1992 and $1.4 billion in 1997 for the Army, Navy, and Air

Force enlisted forces together.[6] The 1997 figure is lower because the military drawdown will reduce personnel inventories between 1992 and 1997.

To estimate the cost savings for all of DoD, we need an estimate for the Marine Corps enlisted force and for officers as well. We make a rough estimate of the cost savings as follows. First, we assume that the percentage cost savings for these forces is equal to a weighted average of the cost savings percentages for the Army, Navy, and Air Force enlisted forces where the weights equal the fraction of total cost (basic pay plus the retirement accrual) that are attributable to the Army, Navy, and Air Force enlisted forces, respectively. This weighted average equals 6 percent.[7] We then applied this 6 percent figure to the total costs for the Marine Corps enlisted force and for officers. For this group, we estimate a cost savings equal to $1.18 billion in 1992 and $.96 billion in 1997. Therefore, we estimate a total cost savings for all active duty military personnel of $3.26 billion in 1992 and $2.4 billion in 1997.

These figures represent a lower bound of the cost savings to the extent that the services may desire a more junior force than the current one. In that case, the increase in active pay necessary to offset the lower expected value of retirement benefits under MFERS would be smaller and the cost savings of the system would therefore be greater. They are also a lower bound to the extent that the retention problems associated with MFERS with a 7 percent pay increase would be addressed with reenlistment bonuses, which would be targeted only to distinct populations and not with a higher pay raise, the option we analyzed. Not all members who would get the pay raise would get the reenlistment bonus, so the version with retention bonuses would imply lower costs.

It was noted in the Summary that the costs of MFERS with a skewed pay raise and REDUX systems are sensitive to the real government discount rate assumption. The accrual charge for either system will fall the higher the discount rate, but REDUX costs will fall more the

---

[6]These cost figures and the ones below are in 1995 dollars.

[7]Although, in principle, the weights will change depending on the calendar year used for the data to calculate the weights, in actuality the weights do not change between 1992 and 1997.

higher the rate because its payments are delayed more. Table 12 shows the cost of the Army enlisted force under real discount rate assumptions of 2 percent, 2.75 percent, and 5 percent.

The savings from MFERS is more than halved when the discount rate is increased to 2.75 percent from 2 percent. MFERS with a skewed pay raise would add 3.8 percent to the cost of the Army enlisted force when the real rate is increased to 5 percent. Using the methods discussed above to extrapolate to the DoD-wide force, MFERS would save about 2.4 percent ($1 billion) in total manpower costs (again based on FY 1997 force levels) if the real discount rate is increased to 2.75 percent. However, it would add about 4.2 percent ($1.7 billion) when the discount rate is increased to 5 percent. The projected savings are therefore very sensitive to the assumption about the real discount rate. (It is important to add that these savings or cost increases are net of the changes in basic pay costs and the retirement accrual. Because of the pay raises required, under an assumption of a 5 percent real rate, MFERS is predicted to add almost 6 percent to annual outlays for the Army enlisted force.)

Although the savings projected for MFERS with a skewed pay raise are modest at best, it must be kept in mind that the analysis is predi-

Table 12

Costs of Alternative Systems Under Different Assumptions About Government Discount Rates
($ billions)

Cost	2 percent		2.75 percent		5 percent	
	REDUX	MFERS	REDUX	MFERS	REDUX	MFERS
Basic pay	9.34	9.97	9.34	10.00	9.34	10.00
Retirement accrual	1.95	.71	1.54	.64	.79	.51
Total	11.29	10.68	10.88	10.64	10.13	10.51
Difference in total cost between REDUX and MFERS		−.61 (−5.4%)		−.24 (−2.2%)		+.38 (+3.8%)

NOTE: MFERS refers to MFERS with a skewed pay raise.

cated on maintaining a force structure similar to the current one. To the extent that MFERS permits the services to move toward different (and for each of them more appropriate) force structures, the potential savings might well be larger. The Air Force might, for example, choose to maintain a relatively senior force, whereas the Army might choose a more junior one (compared to the one produced by REDUX). Within services, different experience structures might be selected for different skill areas. Costs could be further reduced by using bonuses rather than pay increases to maintain retention and productivity in services and skill areas. These and other force management considerations are discussed in Chapter Six.

Chapter Five

RESULTS FOR THE TRANSITION TO THE STEADY STATE

The results of Chapter Four show the predicted effects of MFERS plus a skewed pay raise in the steady state, when all personnel are under the new compensation system. In this chapter, we present results for the transition to the steady state. We consider two cases. In the first case, which we call the grandfathering case, all existing service members are grandfathered under the current system (which we assume, for simplicity, is REDUX) and only new entrants are enrolled into MFERS plus a skewed pay raise. In the second case, which we call the convert case, current members are given the option of converting to MFERS plus a skewed pay raise or remaining in REDUX, while new entrants are automatically enrolled into the new system. Since members who convert would not ordinarily receive credit under MFERS for prior years served, DoD would make double contributions to the thrift savings plan for a time equal to the member's years of service. We show here the predicted retention and cost effects under the two cases. Before presenting the results, we first discuss what we would predict theoretically to occur in transition to the steady state under each case.

THEORETICAL EFFECTS

Case 1: Grandfathering

By construction, both REDUX and MFERS with a skewed pay raise give similar retention patterns in the steady state (see Table 9). As new entrants are enrolled under MFERS with a skewed pay raise, we would not predict any change in their retention incentives relative to REDUX. Since existing members would continue to be covered by

REDUX in the grandfathering case, their retention incentives should also remain unchanged. Therefore, we would not expect any change in retention patterns during the transition to the new steady state in the grandfathering case.

We would expect costs to decline in the transition since MFERS with a skewed pay raise is a less costly system than REDUX. Also, we might expect the largest declines to occur initially since those with few years of service make up the bulk of the enlisted force, and these individuals would be the first to be enrolled into the new and less expensive system. Costs may even fall below their steady-state level for some period because of the skewed pay raise associated with MFERS that gives the larger and more expensive pay raises in the higher grades/years of service, and the smaller and less expensive pay raises in the lower grades/years of service. Although retention incentives are maintained among the more junior personnel because of the promise of large pay increases in the upper grades, DoD does not incur any costs associated with these larger raises until the new entrants reach the more senior grades, a time period that can occur many years into the transition. Although costs may fall below their steady-state level for some time, eventually they would have to rise back to the steady state.

Case 2: Converting Members

As in the grandfathering case, we would expect new entrants who are automatically enrolled in MFERS plus a skewed pay raise to face the same retention incentives as under REDUX because the two systems are designed to produce the same retention patterns in the steady state. Therefore, we would not expect any change in their behavior in the transition in case 2.

However, we would expect a change in the retention incentives of those who convert to the new system because converting members are given a compensation option that neither those in the steady state nor new entrants are given. They have the option of staying in the military under MFERS with a skewed pay raise, staying under REDUX, or leaving. Furthermore, although MFERS with a skewed pay raise provides incentives to leave the service that REDUX does not provide (since it gives a benefit to those who leave prior to YOS 20 and it is portable to federal civil service), it also provides an incentive to stay since it offers a generous pay raise to those in the higher grades, it gives double contributions for a specified time, and individuals must remain in the service for a required period to become vested. Because those who convert do so only because they will be better off, no one is made worse off. Consequently, we would expect retention rates to rise among YOS cohorts whose members are permitted to convert because military service is now more attractive for some personnel.[1]

Given that retention rates under the new and old systems are the same in the steady state, those whose retention rises in transition to the steady state must be individuals with below-average tastes for military service. Thus, at some future date, retention rates among YOS cohorts whose members can convert must drop below the steady-state level because these low-taste individuals will leave the service faster than they would in the steady state. The net result is that among cohorts whose members can convert, retention follows an S-shaped pattern; it initially rises above its steady-state level, drops below the steady-state level, and then returns to the steady-state level.

With voluntary conversion, those who convert would receive double contributions for a time in part to provide credit for past years of service but also to prevent spikes in retention rates. However, even with higher DoD contributions, it would be impossible to prevent the S-shaped retention pattern among converting cohorts in the transition unless DoD could "price discriminate" and offer different contribution rates to different individuals such that all individuals were indifferent between staying and leaving. In other words, for some individuals, it would need to double contributions, while for others it would need to triple contributions. Such a policy of differentiating

---

[1]This rise above the steady-state level is entirely consistent with the fact that retention trends in the steady state are the same under the new and old systems. Retention rates in the steady state are the same for individuals with a marginal taste for military service. But retention rates can differ among inframarginal individuals in transition to the steady state. For example, an inframarginal individual with five years of service but with a relatively low taste for military service might leave service after six years of service in the steady state. But, in the transition to the steady state, such an individual may place a particularly high value on portability of benefits to the federal civil service, choose to convert, and stay in service longer than six years of service to gain vesting and double contributions under MFERS with a skewed pay raise.

among individuals would be prohibitively costly to implement. Therefore, it would not be possible to prevent some variation in retention among cohorts, although the amplitude of the S-curve (i.e., the degree of the spike and valley of the S-curve) would be affected by the overall contribution rate DoD offered in the transition.

Although we can predict the retention patterns over time of YOS cohorts whose members are allowed to convert to the new system, it is more difficult to predict the calendar year-to-year variations in retention because these variations represent an aggregation across YOS cohorts. For example, we would predict that in the initial years of the transition, retention rates would rise as retention rises among cohorts whose members can convert, and the rates of new entrants are unchanged. But over time, the rates of some cohorts will begin to fall while others are still rising. Therefore, the aggregate effect cannot be predicted a priori.

Predicting year-to-year variations in costs under case 2 is also problematic. Obviously, costs fall because the new compensation system is less expensive than REDUX. But costs will rise in the transition as retention among some cohorts rises and will fall in the transition as retention among some cohorts falls. The net effect is unclear but costs could actually fall below the steady-state level for some period of time if the net effect is negative.

RESULTS

Our results are for the Army enlisted force. In general, we find that while retention varies in transition to the steady state under both the grandfathering and convert cases, our model does not predict any large spikes in retention rates. In other words, the transition to the new steady state is predicted to be fairly smooth. The model also predicts that for both cases most of the cost savings associated with transitioning to MFERS with a skewed pay raise occurs fairly early. Furthermore, with the exception of the first few years when there is a large drop in costs, there are no spikes in costs in transition to the new steady state.

Before presenting these results, we show the pattern of conversion in the convert case. Table 13 shows the fraction of the force that is predicted to convert to MFERS plus a skewed pay raise. Overall, 80

Table 13

Predicted Conversion Rates Among the Army Enlisted Force

Percent Converting at:	E-1–E-3	E-4–E-6	E-7–E-9	Total
YOS 1–4	100.0	99.4	0.0	99.7
YOS 5–10	0.0	94.5	100.0	94.6
YOS 11–20	0.0	12.7	11.9	12.4
YOS 21–30	0.0	0.0	0.0	0.0
Total	100.0	79.5	11.2	80.4

percent of the force is predicted to convert to the new system, but this fraction is concentrated among those with 1 to 10 years of service and those in pay grades E-1 to E-6. These individuals are further from the 20-year vesting point under REDUX, and thus have a relatively lower expected value of retirement benefits under REDUX.

Now consider the retention effects in transition to the steady state. In the grandfathering case we predict little change in retention behavior. Figure 1 shows man-years per accession (MPA) for the steady state (represented by a solid line), for the grandfathering case (represented by a dashed line), and for the convert case (represented by dots) from year 1 after the transition to year 30 when all members would be covered by the new system. Relative to the steady state, the model predicts some variation around the steady state in retention in the grandfathering case. This variation is predicted by the model because, as shown in Table 9, the retention patterns predicted in the steady state for MFERS plus a skewed pay raise are close but not identical to those predicted in the steady state for REDUX.[2] Thus, some variation would be expected. However, the variation is not large; at most, the difference in MPA between the steady state and the grandfathering case is on the order of 3 percent.[3]

---

[2]With sufficient manipulation of the skewed pay raise, the retention patterns could be made even closer between REDUX and MFERS with a skewed pay raise.

[3]Small differences in MPA can mask large differences in underlying retention patterns; however, a comparison of conditional retention rates between the steady state and the grandfathering case did not reveal large differences in retention between the two cases.

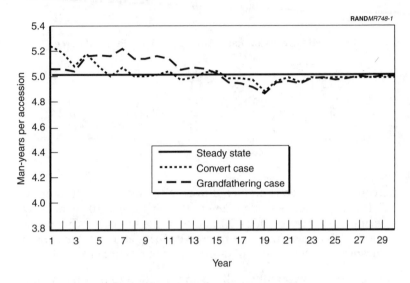

Figure 1—Man-Years per Accession in the Steady State, Convert, and
Grandfathering Cases

For the convert case, we observe a small initial increase in MPA relative to the steady state, as we would expect theoretically, but this increase is not large, a difference of about 5 percent. While a 5 percent difference in MPA can mask a large difference in retention patterns, we do not find large differences in conditional retention rates—another measure of retention—between the steady state and convert case (not shown). Figure 1 shows that MPA in the convert case shows some variation around the steady state over time, including a dip in year 19, but again these variations are not large.

Also as we would expect theoretically, the model predicts an S-shaped pattern in retention for cohorts whose members are allowed to convert. An example is seen in Figure 2, which shows the conditional retention rates over time for the cohort that is at YOS 8 in the first year of the transition. After four years, this cohort has 12 years of service, after eight years, it has 16 years of service, etc. The solid line shows the conditional retention rates that would be predicted in the steady state for cohorts with 8 YOS, 12 YOS, 16 YOS,

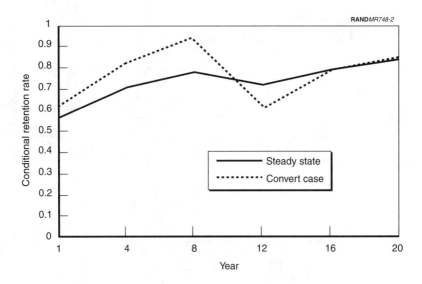

Figure 2—Conditional Retention Rates for the YOS 8 Cohort

etc. The dotted line shows the conditional retention rates predicted in transition to the steady state. The model predicts that retention rises above the steady state as more individuals in the cohort are induced to stay by the option to convert to MFERS plus a skewed pay raise. However, after year 8, retention falls. At year 10, retention falls below the steady state as low-taste individuals who were induced to stay exit at a rate faster than would be predicted in the steady state. At year 16, the cohort retention patterns in the transition to the steady state match those found in the steady state.

The predicted patterns of total costs over the transition to the steady state are shown in Figure 3 for the REDUX steady state (long and short dashes), the MFERS plus skewed pay raise steady state (solid line), the grandfathering case (dashes), and the convert case (dots). Figure 3 is predicated on the assumption of a 2 percent real government discount rate. The time profiles of cost are similar for higher discount rates, but the absolute differences would change. The figure shows the total of basic pay costs, retirement accrual costs, and DoD contribution costs for the Army enlisted force in billions of dol-

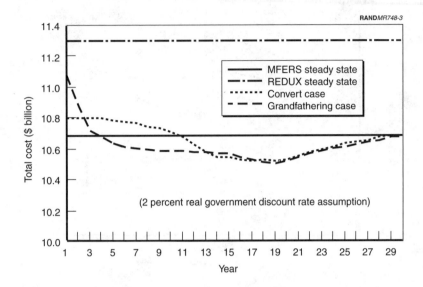

Figure 3—Total Cost in the REDUX Steady State, MFERS Steady State, Grandfathering, and Convert Cases

lars. The model predicts that in the grandfathering case, DoD would realize about 35 percent of the total cost savings associated with moving to MFERS with a skewed pay raise within the first year (a drop from $11.29 billion to $11.18 billion). However, within three years, the model predicts that it would realize all of the cost savings, and in fact costs would drop below the steady state for a period of time.

The drop below the steady state results from a combination of two factors. First, the pay raises given to junior personnel are lower under the skewed pay raise scheme. The larger raises given to more senior personnel are costs that DoD will not incur until the later years of the transition. Second, as noted above, the retention patterns between REDUX and MFERS with a skewed pay raise are close but not exact, resulting in some variation in retention and cost.

In contrast to the grandfathering case, the model predicts that in the convert case about 80 percent of the change associated with moving to MFERS plus a skewed pay raise would be realized in the first year.

More of the change is realized in the first year in the convert case because some of the more senior and more costly personnel are converting to a system that is less expensive. Total cost in the convert case also shows some variation around the new steady state. In part this is because retention is rising among some cohorts while falling among others, and because the REDUX and MFERS steady-state retention patterns are close but not exact.

While we observe some variation around the new steady state in costs for both the convert and grandfathering cases beyond the first three years, the variations are not large. Thus, the model does not predict large spikes (or valleys) in the pattern of costs or retention in transition to the steady state, regardless of which strategy—grandfathering or convert—that DoD chooses. However, the cost differences between MFERS and REDUX would be smaller than those shown in Figure 3 when the real government discount rate is increased to 2.75 percent and MFERS would cost more when the rate is assumed to be 5 percent. Thus, the results are sensitive to the assumption about the real government discount rate.

Chapter Six

OTHER CONSIDERATIONS

The simulation model predicts that MFERS with a skewed pay raise can produce the same general force size and structure as REDUX but at less cost and with higher productivity. However, retention, productivity, and payroll costs are not the only important considerations in determining whether military members should be converted to MFERS. In this chapter, we discuss four other factors to consider in weighing the costs and benefits of moving military members to MFERS. Unfortunately, unlike retention, cost, and to some extent, productivity, these factors are not amenable to measurement. The first factor is the ability of MFERS to increase force management flexibility—a criticism of the current system alluded to in the introduction. The second is the degree to which MFERS would rely on involuntary separation to achieve the desired force structure. The third factor is the portability of benefits under MFERS. The final factor is the likelihood of a more stable force size and structure in the face of business cycles and other changes in the civilian labor market.

FORCE MANAGEMENT FLEXIBILITY

The current military retirement system embodies many of the features one would expect in the compensation system of a hierarchical organization. For example, the delayed benefits effectively skew total compensation toward those reaching the upper ranks, thereby maintaining the motivation and work effort of nonvested personnel. In addition, the generous nature of the benefits for those who become vested induces voluntary separations and helps minimize the organizational influence costs that might attend the separation of senior personnel under less-generous terms.

But, as we noted in the introduction, despite these virtues of the system, a basic criticism of the 20-year system regards force management flexibility. At a very general level, the retirement system creates the implicit contract problem discussed earlier. The prospect of retirement after 20 years is a delayed "carrot" that induces personnel to invest in military-specific job skills, to accept onerous or hazardous assignments, and generally to exert work effort early in their careers. Individuals, of course, will not make such investments without a good chance that they will pay off. Therefore, beyond a certain career point involuntary separations would appear capricious and would adversely affect the incentive scheme. The services are understandably reluctant to separate mid-career personnel for fear of how such separations will affect the behavior of more junior personnel. The 20-year system creates a kind of implicit contract or guarantee of tenure to mid-careerists and, arguably, has the effect of inducing the services to "demand" more mid-careerists than they might under a different system.

That the terms of separation affect force management practices is illustrated by the Army's Qualitative Management Program (QMP). Under QMP, a board of senior enlisted personnel meets annually to select for involuntary separation approximately 2 to 3 percent of the lowest performers in grades E-5 through E-9. However, the board selects for separation only those who are retirement-eligible. Recognizing the financial costs imposed on those who have not yet qualified for retirement benefits, the board selects for separation only those who would not be excessively financially penalized by involuntary separation. It is likely that all of the services have carried to the 20-year point many personnel who would have been separated earlier under a different system.

Retention trends during the All-Volunteer Force (AVF) era have compounded the problem. Higher first-term retention in the AVF meant larger flows into the career force and more personnel competing for promotion to the upper ranks. The implicit contract to mid-careerists limited the services' ability to control flows of mid-career personnel and reduced promotion opportunities for younger personnel. Overall, the fraction of the enlisted forces with more than 10 years of service rose by about 25 percent over the 1974–1989 period, with the largest seniority increases in the Army (43 percent) and the Marine Corps (49 percent). In fact, the Navy and Air Force experienced little increase in the fractions of their enlisted forces with more than 10 years of service.

Although increased enlisted seniority might theoretically be welcomed on the ground that more experienced forces are more productive, it is important to note that the seniority growth occurred in the two services that profess the most need for youth and vigor in their enlisted forces. The seniority growth raised serious questions about cost and made evident the services' inability to effectively manage their senior enlisted forces. After considerable pressure from the Office of the Secretary of Defense (OSD), in 1990 the services began applying more-stringent high-year-of-tenure rules to their enlisted forces. But these more-stringent rules affected relatively few personnel who were not retirement-eligible. It was the large force reductions that began after 1990 that forced the services to seriously consider separating significant numbers of mid-career personnel. At first, the services wanted to reduce their strengths by cutting accessions, but the implications of this policy for the future force structure soon became clear. It was only after the implementation of the Voluntary Separation Incentive (VSI) and the Special Separation Benefit (SSB) schemes that the services agreed to reductions in the mid-career force. These temporary separation payment schemes expired in 1995, but the experience with them so far illustrates how force management practices would change with different terms of separation.

A related point is that when the quality of entering cohorts varies significantly, the retirement system compounds the difficulty of managing quality flows through the force. Cohorts entering the Army in the late 1970s were of poorer quality than later cohorts. High retention of these cohorts as they entered their second decade of service clogged the mid-ranks and increased the difficulty of retaining and advancing the higher-aptitude personnel in the later cohorts. The separation tools offered by the drawdown program have enabled the Army to selectively separate the less-able personnel, something it could not have done before.

The 20-year system poses difficulties at more detailed levels. The system is identical for all (active) members regardless of occupation or service and regardless of whether the individual is an officer or a member of the enlisted force. Yet occupations, services, and officer and enlisted roles are obviously different. One important way that occupations differ is in their desired experience profiles. In some occupations, notably combat arms skills, a youthful experience profile is required. In others, youth and vigor are not primary job requirements, and high training costs and/or a big payoff to job experience (such as with doctors and nurses) argue for longer than 20-year military careers. But as shown in Asch and Warner (1994b), the system produces similar force profiles across the broad spectrum of occupations. Thus, force managers seem to have little flexibility in shaping or controlling the experience profiles of the various occupations (or services).

In terms of improving force management flexibility, would MFERS be an improvement over the current system? MFERS eliminates vesting at the 20-year point and reduces the value of retirement benefits at YOS 20. Therefore, there is not the pull toward a 20-year career for mid-careerists that holds in the current retirement system. Furthermore, since MFERS allows those who leave prior to YOS 20 to get some benefits, the services are likely to be more willing to involuntarily separate personnel whom they would not separate under the current system.

However, other factors come into play here. Although they may be more willing, there are two reasons why the services would likely continue under MFERS to act as if members are serving under an implicit contract. First, the skewed pay raise that must accompany MFERS for MFERS to be an improvement over REDUX replaces the retirement system as the source of deferred compensation and therefore the source of productivity and retention incentives. The skewed pay raise gives higher raises to those in higher grades, but those who reach the higher grades are those with more years of service. The services are likely to be unwilling to involuntarily separate those in their mid-career who are "due" large raises in their later years of service. An implicit contract may be formed because of MFERS' large deferred pay raises. Similarly, an implicit contract may be formed if the services use skewed retention bonuses rather than skewed pay raises to address retention problems that arise with

MFERS plus a 7 percent across-the-board pay raise. Just as in the skewed pay raise case, if members feel that they are "owed" the retention bonuses coming in the higher grades or years of service, the services may be extremely reluctant to separate them.

Second, the present value of the retirement benefit that a member would get under MFERS if he or she separated prior to YOS 20 is relatively small (less than one year of base pay[1]), as shown in Table 6. This separation benefit is unlikely to fully compensate many members for the second-career loss associated with transitioning to the civilian sector. Some personnel whom the services would prefer to leave will opt to stay given these relatively small benefits under MFERS. If the services are unwilling to impose a financial loss on these members, even though the financial loss is smaller under MFERS than under the current system, they will continue to act as if members are serving under an implicit contract. Therefore, although MFERS addresses the implicit contract problem associated with the current military retirement system, it creates its own implicit contract problem, so that the amount of force management flexibility it would afford, especially for mid-careerists, is questionable.

Like the current system, MFERS with a skewed pay raise would be identical for all members regardless of occupation or service and regardless of whether the individual is an officer or a member of the enlisted force. Put differently, it would also be a "one-size-fits-all" system, subject to the same difficulties of force management flexibility at the more detailed level as the current system. On the other hand, if MFERS is coupled with a system of retention bonuses that could be targeted to distinct populations, then MFERS could address some of the force management flexibility problems associated with the current system. A compensation system that includes MFERS could be designed that allows the services to achieve varying experience profiles across occupations or personnel types.

---

[1]In comparison, the SSB program (the lump-sum separation pay program being used by the services to facilitate the drawdown) pays 1.5 times base pay.

INVOLUNTARY SEPARATION

As discussed in the overview of the theoretical model in Chapter Four, retired pay can be used to induce voluntary separations of senior personnel. In a hierarchy without lateral entry, separation of more senior personnel is necessary to maintain promotion opportunities and provide retention and productivity incentives to more junior personnel. The military could also induce the separation of more senior personnel through involuntary means but, as discussed in Chapter Four, involuntary separations create organizational influence costs. For example, involuntary separations lower morale, which adversely affects retention of more junior personnel and possibly recruiting. To restore retention and recruiting, pay would need to be raised. The cost of the pay increase is an organizational influence cost.

In addition, when the services rely on involuntary separations, personnel are likely to engage in practices to loosen the policy—in the hope of coercing the services to make fewer involuntary separations. If the services respond to this pressure, it will result in a superannuated or older force and, to the extent that the services desire "youth and vigor," a less productive force. The lower productivity of the force is another organizational influence cost. If the services paid separation pay to trim the force, the cost of separation pay would be an organizational influence cost.

A compensation system that defers compensation in the form of a retirement benefit that provides an immediate benefit upon separation saves these organizational influence costs. The current military system is such a system. A distinct advantage of the current system is that the separation of more senior personnel is voluntary.

A disadvantage of a front-loaded compensation system such as MFERS is that it relies upon involuntary separation. Under MFERS, the military would rely more on HYT rules and other forms of involuntary separation to maintain a youthful force. Table 14 shows that in the absence of HYT rules, our simulation model predicts that MFERS with a skewed pay raise would generate a more senior force than would REDUX. For example, comparing the top and bottom panels, survival to YOS 30 would rise to 1.9 percent under REDUX but would rise to 4.1 percent under MFERS with a skewed pay raise. The

Table 14

Predicted Effects of REDUX and MFERS with a Skewed Pay Raise without HYT Rules

	I. REDUX					
	Grade-by-YOS Distribution				Survival to	
YOS	E-1–E-3	E-4–E-6	E-7–E-9	Total	Start of	
YOS 1–4	.288	.187	.000	.475	YOS 5	.327
YOS 5–10	.006	.230	.003	.239	YOS 10	.168
YOS 11–20	.001	.138	.065	.204	YOS 20	.110
YOS 21–30	.000	.021	.061	.082	YOS 30	.019
Total	.295	.576	.129			
	II. MFERS with a Skewed Pay Raise					
	Grade-by-YOS Distribution				Survival to	
YOS	E-1–E-3	E-4–E-6	E-7–E-9	Total	Start of	
YOS 1–4	.282	.187	.000	.469	YOS 5	.350
YOS 5–10	.006	.238	.003	.247	YOS 10	.174
YOS 11–20	.000	.118	.065	.183	YOS 20	.095
YOS 21–30	.000	.000	.076	.076	YOS 30	.041
Total	.288	.543	.144			

fraction of the force in YOS 20 to YOS 30 would rise to 8.2 percent and 10.0 percent under each system, respectively. The implication of the results in Table 14 is that to achieve the same force profile as the current system, MFERS with a skewed pay raise must subject more individuals to HYT rules—to involuntary separation.

Although those who are involuntarily separated under MFERS would receive some benefit—MFERS vests personnel as early as YOS 3—the services are still likely to face more organizational influence costs under MFERS than they do under the current system. The reason is that the size of the retirement benefit under MFERS is relatively small, as noted above, so that some of those who would leave voluntarily under the current system will not leave voluntarily under MFERS. Of course, the services may desire to retain some of these individuals, particularly those for whom a longer than 20-year career is appropriate. But to maintain the structure of the force, other individuals will need to be involuntarily separated, a policy that will give rise to organizational influence costs. How large the organizational influence costs will be under MFERS is unclear. The problem of involuntary separations in the senior YOS may be self-limiting to some extent since retirement from service is mandatory at YOS 30 (except by special waiver), and many members will begin their search for a second career around YOS 25. Their incentives will depend critically on how civilian opportunities change from age 45 to age 50 and on the variance in civilian opportunities. Still, since the cost differential between MFERS with a pay raise and the current system is not large, the organizational influence costs associated with involuntary separation do not need to be large before they eliminate the MFERS cost advantage.

The organizational influence costs under MFERS will depend on whether MFERS is coupled with a skewed pay raise, as analyzed in Chapter Four, or with a 7 percent pay increase and a system of retention bonuses. They are likely to be larger in the former than in the latter case. The services can induce some voluntary separations in the latter case that would not occur in the former case by simply turning the retention bonuses off at the appropriate time. Thus, some of the separations that would occur involuntarily under MFERS with a skewed pay raise would occur voluntarily under MFERS with a system of retention bonuses. Organizational influence costs are likely to be smaller then because fewer separations occur involuntarily when MFERS is coupled with retention bonuses.

On the other hand, bonuses operate only at the margin. When the services turn off the bonuses, some personnel will leave voluntarily (those at the margin), but those with a strong taste for military service will stay despite the reduction in bonus income. The services will have to eventually involuntarily separate these high-taste individuals. Involuntary separations will be necessary even when MFERS is coupled with retention bonuses.

PORTABILITY

There is limited evidence on the number of veterans who transition into the civil service upon separating from the military. Anecdotal evidence suggests the number is high. DMDC data suggest that some 10 to 15 percent of military retirees—those who reach at least

20 years of service—enter the civil service.[2] Because of these numbers, integration of the military and civil service retirement systems would seem particularly advantageous. If the systems were integrated, members could transfer their fund accumulations to the new system with no penalty when they changed jobs. An obvious advantage of converting military members to MFERS is that it would allow the integration of MFERS with FERS—the military with the civil service systems.

A related advantage of converting military members to MFERS is that FERS is already in place for federal employees. A political barrier to the implementation of any new military system is that the system is untried. MFERS is less subject to this problem than other proposed systems.

FORCE STABILITY

Because the military does not allow lateral entry into the upper grades, the services must grow their career forces. Those in the career force who leave too early either create an undesirable vacancy or necessitate quicker-than-desired promotions from the lower ranks to fill the vacancy. Thus, premature losses from the career force impose a cost on the services.

Why do premature losses occur? One reason is that there may be random fluctuations in the civilian labor market that make civilian employment more attractive than anticipated. Similarly, higher-than-expected retention may occur during business downturns that make civilian employment less attractive.

The number of premature losses (or amount of unexpectedly high retention) will depend on the military compensation system. Evidence shows that turnover is reduced when some compensation is deferred, such as in the form of a pension (see for example, Mitchell [1982]). A distinct advantage of the current military retirement system is that a large fraction of military compensation takes the form of deferred retired pay. The expected value of retired pay is

---

[2]Specifically, 14.8 percent of those who retired from the military in 1983 were in federal civil service in 1995, and 9.2 percent of those retiring in 1989 were in federal civil service in 1995.

greatest for those in the senior grades. The current retirement system therefore buffers the services from unexpected personnel losses in the senior force resulting from random fluctuations in the civilian labor market. Empirical studies of first- and second-term enlisted retention indicate that retention for more junior personnel is sensitive to civilian opportunities. Thus, stability of the current force is achieved primarily for the more senior personnel. Of course, a cost of having this greater stability is that during economic downturns, retention among this group may be greater than desired.

MFERS front-loads compensation to a greater extent than does the current system and so is less likely to buffer the services against retention fluctuations. On the other hand, when MFERS is coupled with an across-the-board pay raise, it might give added protection against retention fluctuations among more junior personnel. Also, when MFERS is coupled with a skewed pay raise, much of compensation would still be deferred in the form of large raises in the senior grades. Thus, the system would still tie senior personnel to the military in the face of random economic fluctuations as well as junior personnel who anticipate remaining in the service. As discussed above, the large raises in the senior grades create an implicit contract that is likely to buffer the services from fluctuations in retention. It is therefore unclear whether MFERS—with either skewed pay or an across-the-board pay raise and skewed bonuses—creates more or less force stability relative to the current retirement system in the face of changing civilian opportunities.

Chapter Seven

CONCLUSIONS AND POLICY OPTIONS

We have compared the current military retirement system as of August 1986 (commonly known as REDUX) with a proposed alternative (which we call MFERS) patterned after the retirement system for federal employees. The comparison included productivity, cost, implications for individual service members, and implications for force management. MFERS attempts to correct three alleged deficiencies with the current system—its unfairness to mid-career personnel who upon separation leave without retirement benefits, the lack of portability of the current system to other retirement systems including the federal system, and its cost. Although we recognize the first two potential deficiencies, they are not the focus of our analysis. The central focus of our analysis has been to determine how MFERS would affect both the cost and the productivity of military forces and whether MFERS would improve or hinder force management. We summarize our conclusions below.

MFERS coupled with an increase in basic pay that just maintains existing take-home pay would result in reductions in retention, experience, and productivity. These reductions occur because MFERS represents a significant decrease in lifetime compensation for personnel with long military careers. This decrease is not offset by increased benefits to those who separate before the 20-year mark. MFERS would need to be coupled with an increase in active duty pay to maintain the experience level and productivity of the REDUX force. Furthermore, the pay increases would need to be skewed—targeted to the higher ranks—to maintain productivity, because MFERS would undo the skewing in the current system.

Our analysis indicates that when coupled with a skewed pay increase that maintains the same general force size and structure as the current system, MFERS would raise productivity. It would do so by providing a greater incentive to supply effort and a higher retention of more-able personnel. Measured on grounds of productivity, MFERS appears to be an improvement over the current system.

How MFERS would affect manpower costs depends on the assumed government discount rate. Manpower costs consist of active duty pay outlays plus the accrual charge for future retirement liabilities. When the assumed government discount rate is 2 percent, MFERS is estimated to reduce total manpower costs by about 6 percent and produce a total force savings of about $2.4 billion. Savings are produced because MFERS reduces the military retirement accrual charge by more than it increases active duty pay outlays. Savings decline when the discount rate is increased; in fact, MFERS is estimated to increase total manpower costs at a real government discount rate of 5 percent. Therefore, although MFERS appears to improve upon REDUX on grounds of both productivity and cost at low discount rates, the case for MFERS is less clear at higher discount rates.

It has been argued that because the current system places such a high portion of compensation in the form of retirement benefits, it hampers force management and creates inflexibilities that could be avoided with a more front-loaded compensation system. In particular, it is argued that a more up-front system would (1) have more flexibility to target pay to services and skills with retention problems and (2) avoid the "implicit contract" problem of the services retaining mid-career personnel who are marginal performers but who are not yet vested in the 20-year retirement system. Although there may be merit to these criticisms, MFERS with skewed increases in active basic pay is not likely to solve them.

First, while in theory MFERS with a skewed pay raise addresses the implicit contract problem produced by the current system in that it offers a benefit to those who leave before the 20-year point, the size of these benefits is small, and the large raises associated with the higher grades may cause the services to retain marginal performers who are "due" a high raise. Retention of some of these individuals may be desirable if they are in occupational areas in which a long career is desirable. But for others, the high raises in the higher grades may create the same implicit contract problems found in the current retirement system.

Second, like the current system, MFERS with a skewed pay raise would be a one-size-fits-all system. Since the system cannot be tailored to meet the unique needs of specific populations such as occupations and services, it is likely to hamper force management flexibility at the micro level.

Third, MFERS with a skewed pay raise would rely to a greater extent on involuntary separations to maintain the youth and vigor of the force because it provides compensation more in the form of active duty pay and less in the form of retired pay—a form of compensation that induces voluntary separations at the appropriate time. Involuntary separations create ex post regret. Although service members were willing to enter the service knowing that most of their compensation would be in the form of active pay and less in the form of retired/separation pay, and that they could be involuntarily separated, once they leave they regret having a compensation system that pays little separation pay and that involuntarily separates members. The ex post regret associated with involuntary separations creates organizational influence costs. Involuntary separations hurt morale, and must be offset by costly pay raises to maintain recruiting and retention outcomes. In addition, if the services choose to relax the involuntary separation policy, the force becomes older, with less youth and vigor, and the productivity of the force declines, another costly outcome. These organizational influence costs are difficult to measure but they may well swamp the cost advantage of MFERS with a skewed pay raise predicted by our empirical model.

MFERS WITH RETENTION BONUSES

An alternative to MFERS with a skewed pay raise is a system that would offer MFERS plus a 7 percent across-the-board pay raise to offset mandatory contributions and their tax consequences under the basic retirement plan, and a system of retention bonuses intended to solve any retention problems that arose. The bonuses could be "turned off" to induce members to leave the service. We could not analyze this system because our model is not occupation-specific and could not easily accommodate retention bonuses. Still, the plan we analyzed is similar in that it would also offer MFERS and up-front compensation in the form of either active pay or active pay/retention bonuses.

Would our conclusions regarding MFERS with retention bonuses likely differ from the ones derived for MFERS plus a skewed pay raise? Our answer is yes, but probably not by much qualitatively. To the extent that the retention bonuses were skewed (higher in higher grades), MFERS with a 7 percent pay raise and a skewed system of bonuses is likely to increase productivity and reduce costs relative to the current system. This system will be portable to the civil service. Thus, MFERS with retention bonuses is likely to create the same three advantages over the current system that were found in MFERS with a skewed pay raise. In fact, the version with bonuses would probably cost less than the version with a skewed pay raise if the bonuses were targeted to only certain populations. Further, since the retention bonuses can be targeted, MFERS with retention bonuses is not a one-size-fits-all plan. Thus, one would expect more force management flexibility at the micro level under this plan than under MFERS with a skewed pay raise.

With the exception of the one-size-fits-all problem, MFERS with retention bonuses would likely be subject to the same disadvantages as MFERS with a skewed pay raise. Both systems would probably create an implicit contract if the retention bonuses and pay were skewed to maintain effort and ability sorting incentives. The services are likely to be reluctant to involuntarily separate any member who is "owed" a big pay raise or retention bonus in a senior grade. Second, both systems would likely involve involuntary separations. While it is true that turning off retention bonuses will induce some voluntary separations, they only operate at the margin. Those who have a strong taste for military service will stay despite the lack of a bonus, and it is these individuals who must be involuntarily separated to maintain the youth and vigor of the force. Thus, MFERS with a 7 percent pay raise and a system of retention bonuses will also generate (potentially large) organizational influence costs.

POLICY OPTIONS

Should MFERS, either with a skewed pay raise or with retention bonuses, be adopted? Much depends on how one weighs the advantages and disadvantages of these systems. A priori, we cannot make a judgment.

However, our analysis indicated that there is another MFERS alternative. The ideal retirement system alternative would be an improvement over the current system on (at least) all the dimensions analyzed: retention, productivity, cost, force management flexibility, portability, voluntary separations, and political acceptability. MFERS with a skewed pay raise *and* a system of separation pay would come very close to this ideal system. In other words, by adding separation pay, the military would have a compensation system that would have the advantages of MFERS with a skewed pay raise but would also address many of its disadvantages. It would have the advantages of MFERS with retention bonuses, but address its disadvantages. The separation pay system would be a generalization of the current involuntary separation pay program. Specifically, the separation payment would equal spm*YOS*final pay, where spm is the separation pay multiplier.

To show how this system would produce the cost and productivity advantages of MFERS with a skewed pay raise (but no separation pay), Table 15 gives the predicted effects of MFERS plus a skewed pay raise plus separation pay when we assume an spm and degree of skewness that produces the same general force size and structure as REDUX.[1] The results in Table 15 compared with those in Tables 10 and 11 show that on these dimensions, MFERS plus a skewed pay raise plus separation pay would be an improvement over the current system (although not to the same degree as MFERS with a skewed pay raise but no separation pay).[2] This system would retain the portability and political expediency advantages of MFERS.

---

[1]Specifically, we set spm =.1, assume that all members who have 10 or more years of service are eligible for separation pay, and assume that the skewed pay raises are as follows: E-1 to E-4, 0 percent; E-5, 4 percent; E-6, 8 percent; E-7, 12 percent; E-8, 16 percent; and E-9, 20 percent. Comparing these skewness assumptions with those in Table 7 shows that the pay raise and degree of skewness that are necessary to maintain the force size and structure are less when MFERS is also coupled with separation pay. These assumptions are not unique. The same force size and structure could be roughly achieved with a higher spm and a smaller set of raises or a smaller spm and a larger set of raises.

[2]It should be noted that our analysis (not shown) indicates that for this system to be an improvement over REDUX, the pay raise must be skewed. We find that MFERS

Table 15

**Predicted Effects of MFERS Plus Skewed Pay Raise
Plus Separation Pay**

Productivity Measure	
Average effort	3.64
Average E-9 ability	.112
Cost Measure ($billions)	
Basic pay	9.61
Retirement accrual	1.32
Total	10.93

NOTE: The cost analysis assumes a 2 percent real
government discount rate.

The main advantage of adding separation pay to MFERS is that the
separation pay could be targeted to specific groups and would reduce the organizational influence costs associated with involuntary
separation (separating personnel would be eligible for separation
pay). Since the services would have a tool they could use to ease the
separation of personnel (much like the VSI/SSB program used during
the drawdown), they would be more willing to separate personnel
who are "due" large raises in the senior grades. Therefore, adding
separation pay addresses the force management flexibility disadvantages and involuntary separation disadvantages that would likely exist under MFERS with a skewed pay raise (but no separation pay) or
with retention bonuses.[3]

There is a potential drawback to MFERS with a skewed raise and separation pay. The separation pay might be operated like a bonus program that is subject to frequent changes by personnel managers or
budgeteers. Frequent changes in separation pay would create uncertainty about benefits and have adverse effects on behavior.

---

with separation pay with an across-the-board pay raise of 7 percent would produce
about the same force size and structure as REDUX, but would produce lower productivity, that is, a lower quality force.

[3]This system—MFERS plus a skewed raise plus separation pay—is very similar to the
three-part retirement system analyzed in Asch and Warner (1994b), which was also
shown to be an improvement over the current system. The three-part plan consists of
an old-age annuity vested at 10 years of service, a skewed pay raise, and separation
pay.

Therefore, once the separation pay scheme is in place, the formula and target populations should be changed rarely.

In conclusion, our analysis suggests that the current military retire- ment system can be improved. We show here that MFERS coupled with either a skewed pay raise or an across-the-board 7 percent pay raise plus a set of retention bonuses would be better than the current system on some dimensions but not others. Therefore, what system should be adopted depends on how the systems' advantages and disadvantages are weighed. It is our view that MFERS with a skewed pay raise *and* a system of separation pay would be better than either of the alternatives or the current system.

REFERENCES

Asch, Beth J., and John T. Warner, *A Theory of Military Compensation and Personnel Policy*, RAND, MR-439-OSD, 1994a.

Asch, Beth J., and John T. Warner, *A Policy Analysis of Alternative Retirement Systems*, RAND, MR-465-OSD, 1994b.

Black, Matthew, *Personal Discount Rates: Estimates for the Military Population*, Arlington, VA: SRA Corporation, 1983.

Black, Matthew, Robert Moffitt, and John T. Warner, "The Dynamics of Job Separation: The Case of Federal Employees," *Journal of Applied Econometrics*, Vol. 5, July/September 1990, pp. 245–262.

Borjas, George J., and Finis Welch, "The Postservice Earnings of Military Retirees," in Curtis L. Gilroy (ed.), *Army Manpower Economics*, Boulder, CO: Westview Press, pp. 295–312.

Career Compensation for the Uniformed Services, U.S. Government Printing Office, December 1948.

Daula, Thomas, and Robert Moffitt, "Estimating Dynamic Models of Quit Behavior: The Case of Military Reenlistment," *Journal of Labor Economics*, Vol. 13, No. 3, 1995.

Defense Manpower Commission, *Defense Manpower: The Keystone to National Security*, Report to the President and the Congress, Washington, DC, April 1976.

Final Report of the President's Private-Sector Commission on Government Management (the Grace Commission), U.S. Government Printing Office, March 1985.

Gansler, Jacques S., *Affording Defense*, Cambridge, MA: MIT Press, 1989.

Gilman, Harry J., *Determinants of Personal Discount Rates: An Empirical Examination of the Pattern of Voluntary Contributions of Employees in Four Firms*, Memorandum 76-0984-30, Alexandria, VA: The Center for Naval Analysis, 1976.

Gotz, Glenn, and John J. McCall, *A Dynamic Retention Model for Air Force Officers: Theory and Estimates*, RAND, R-3028-AF, 1984.

Lawrence, Emily C., "Poverty and Rates of Time Preference: Evidence from Panel Data," *Journal of Political Economy*, Vol. 99, 1991.

Lazear, Edward P., and Sherwin Rosen, "Rank-Order Tournaments as Optimum Labor Contracts," *Journal of Political Economy*, Vol. 89, No. 51, 1981, pp. 841–864.

Milgrom, Paul R., "Employment Contracts, Influence Activities, and Efficient Organization Design," *Journal of Political Economy*, Vol. 96, No. 1, 1988, pp. 42–60.

Mitchell, Olivia, "Fringe Benefits and Labor Mobility," *Journal of Human Resources*, Vol. 17, 1982, pp. 286–298.

Modernizing Military Pay, Report of the First Quadrennial Review of Military Compensation, Volume I, *Active Duty Pay*, Department of Defense, November 1967.

Modernizing Military Pay, Report of the First Quadrennial Review of Military Compensation, Volume IV, *The Military Estate Program*, Department of Defense, January 1969.

Press, William, Saul Teukolsky, William Vetterling, and Brian Flannery, *Numerical Recipes in Fortran, 2nd ed.*, New York: Cambridge University Press, 1992.

Report of the President's Commission on Military Compensation, U.S. Government Printing Office, April 1978.

Report of the Seventh Quadrennial Review of Military Compensation, Department of Defense, August 1992.

Rosen, Sherwin, "The Military as an Internal Labor Market: Some Allocation, Productivity, and Incentive Problems," *Social Science Quarterly,* Vol. 73, No. 2, June 1992.

Salop, Joanne, and Steven Salop, "Self-Selection and Turnover in the Labor Market," *Quarterly Journal of Economics,* Vol. 91, 1976, pp. 619–628.

Smith, D. Alton, Stephen D. Sylwester, and Christine M. Villa, "Army Reenlistment Models," in David K. Horne, Curtis L. Gilroy, and D. Alton Smith (eds.), *Military Compensation and Personnel Retention: Models and Evidence,* Alexandria VA: Army Research Institute for the Behavioral and Social Sciences, 1991.

Warner, John T., and Saul Pleeter, "The Personal Discount Rate: Evidence from Military Downsizing Programs," draft paper, Clemson University, South Carolina, December 1995.

Willis, Robert J., and Sherwin Rosen, "Education and Self-Selection," *Journal of Political Economy,* Vol. 87, October 1979, pp. S7–S36.